GEOTECHNICAL ENGINEERING AND SOIL SCIENCE

GEOTECHNICAL ENGINEERING AND SOIL SCIENCE

K S Murty
M M Subarao

Kruger Brentt

Publishers

2023

Kruger Brentt Publishers UK. LTD.
Company Number 9728962

Regd. Office: 68 St Margarets Road, Edgware, Middlesex HA8 9UU

© 2023 AUTHORS
ISBN: 9781787150591

For information on all our publications visit our website at http://krugerbrentt.com/

PREFACE

Geotechnical engineering is a practice that relates to the engineering behaviour of the earth and its materials. As a branch of civil engineering, it is of great importance to construction activities taking place on the surface or within the ground, as well as to mining, coastal, drilling and other disciplines. Geotechnical engineering is important because it helps prevent complications before they happen. Without the advanced calculations and testing provided by a geotech, buildings could see significant damage after an earthquake, slope stability shifting, ongoing settlement, or other effects. For quality control of construction materials, such as sand, gravel, or crushed rocks, an engineer with a geological background is needed. The knowledge of the nature of the rocks in a specific area is necessary for tunneling and determining the stability of cuts and slopes. Geological maps also help in planning projects. If geological features, such as faults, joints, beds, folds, or channels are encountered, suitable remedies should be incorporated. Geological maps provide information regarding the structural disposition of rock types in a proposed area. Topographical maps are essential for understanding the advantages and disadvantages of all possible sites. Hydrological maps provide information regarding the distribution of surface water channels and the occurrence and depth contour of ground water. Knowledge of ground water is necessary for excavation works. Understanding soil erosion transportation and deposition by surface water helps in soil conservation, river control, and coastal works. In geologically-sensitive areas, such as coastal belts and seismic zones, knowledge of the geological history of the area is very important.

Geotechnical engineering is a practice that relates to the engineering behaviour of the earth and its materials. As a branch of civil engineering it is of great importance to construction activities taking place on the surface or within the ground, as well as to mining, coastal, drilling and other disciplines. Despite having considerable overlap, geotechnical engineering differs from engineering geology in that it is a speciality of engineering, whereas the latter is a speciality of geology. Geotechnical engineers are responsible for evaluating subsurface and soil conditions and materials, using the principles of soil and rock mechanics. They are commonly appointed as consultants on construction projects. Engineers also examine environmental issues such as flood plains and water tables. By doing so, they are able to determine whether a particular site is suitable for a proposed project, and can inform the engineering design process with regard to how ground conditions can be made safe and effective for construction. From geotechnical surveys, engineers will be able to evaluate the stability of the ground, including any slopes and soil deposits, assess any risks and/or contamination, and help to determine the types of foundations and earthworks that will be required. The potential for hazards such as landslides, earthquakes and other seismic activity can also be assessed. Geotechnical engineers can be involved in 'ground improvement', in which soil is treated through a variety of different techniques to improve strength, stiffness, and/or permeability. Geotechnical engineering is also important in coastal and ocean engineering, in relation to building wharves, jetties, marinas and coastal defences, as well as foundation and anchor systems for offshore structures such as oil rig platforms. Engineers may also work on embankments, tunnels, channels, reservoirs, irrigation systems, and so on.

The present book contains fifteen chapters covering all related disciplines. These chapters include Introduction, geotechnical properties of soil, permeability and seepage, compaction, stress distribution in soils, shear strength of soils, stability of slopes, geo-environmental engineering, weathering - soil formation factors, processes and components, soil water retention: movement of soil water, infiltration, percolation, permeability and drainage, soil organic matter, soil thermal properties, layer silicate clays - genesis and classification, soil and the carbon cycle and soil colloids. This book is designed to provide in-depth information about Geotechnical Engineering and Soil Science. This book will serve as a valuable resource for graduate and postgraduate students' teachers, researchers, geotechnical, mining and civil engineers for graduate and postgraduate students and academics.

We are grateful to all those persons as well as various books, manuals, periodicals, magazines, journals etc. that helped in the preparation of this book. In spite of the best efforts, it is possible that some errors may have occurred into the compilation and editing of the book. Further queries, constructive suggestions and criticisms for the improvement of the book are always welcome and shall be thankfully acknowledged.

K S Murty

M M Subarao

CONTENTS

1

INTRODUCTION

Geotechnical Engineering has been - throughout the years - an often overlooked but nonetheless integral part of human endeavors towards progress. Everything around us is supported by soil or rock and Geotechnical Engineering is responsible for that. Anything that is not supported by soil or rock, either floats, flies or falls down.

Geotechnical Engineering is the science that explains mechanics of soil and rock and its applications to the development of human kind. It includes, without being limited to, the analysis, design and construction of foundations, slopes, retaining structures, embankments, roadways, tunnels, levees, wharves, landfills and other systems that are made of or are supported by soil or rock.

As the societal and economic systems develop and become more complex, the quest for a more sustainable and resilient built environment is increasingly pertinent. To that end, Geotechnical Engineers, being able to affect every project from its very early stages, play a critical role.

Civil engineering is the professional engineering discipline that deals with the design, construction and maintenance of public and private infrastructure within the natural environment. Geotechnical engineering is a discipline within civil engineering that focuses on the behavior of natural geological materials in engineered systems. Geotechnical engineers recognize that soil and rock are the cheapest and most abundant building materials on Earth, and consequently play a major role in the construction and performance of every type of civil engineering structure.

To be successful in the field of geotechnical engineering, students should have broad exposure to civil engineering, with advanced knowledge and coursework in geology, soil and rock mechanics, slope stability, foundation engineering and computational mechanics.

With this in mind, we have designed our program to prepare students for success. The Geotechnical Engineering program at the University of Delaware offers opportunities for advanced study and research in:

- ⊙ Soil and rock mechanics
- ⊙ Soil-structure interaction
- ⊙ Constitutive modeling
- ⊙ Computational geomechanics
- ⊙ Foundation and earth structures engineering
- ⊙ Ground improvement
- ⊙ Slope stability and landslide stabilization
- ⊙ Liquefaction of soils and earthquake engineering
- ⊙ Laboratory characterization of geomaterials and soil reinforcement
- ⊙ Environmental geotechnics

ADVANCES IN GEOTECHNICAL ENGINEERING

Geotechnical engineering is the branch of civil engineering that studies soils, structures, or structural components residing on or inside these soils. The mechanics of soils provides the necessary physical attributes to be used to understand the mechanisms that play a key role in what happens to soils under external effects, such as those induced by structural loads or changes in the water content. Thus, soil mechanics is mainly interested in the deformation characteristics of soils such that failure loads can be predicted more accurately. Therefore, it is more convenient and perhaps more accurate to think of the problem as a soil-structure system which is composed of a porous soil medium (as a soil profile with number of layers) and a foundation or another geotechnical engineering structure. As for the design of such soil-structure systems, settlement and the ultimate bearing capacity become the two major criteria that must be satisfied to withstand any external static or dynamic load. Once such stability criteria are set, it is expected that any soil-structure system achieves these requirements during the lifetime of the project.

Geotechnical engineering consists of the collection of subfields in a wide range of practical problems related to soils, rocks, slopes, foundations, walls, etc. While the principles of soil mechanics is the ultimate cookbook, recipe for each problem changes, and the engineers who are the chefs of the practice must be able to deduce the most edible and desired cuisine (i.e., engineering solution) employing their ultimate material, mathematics. As the current technology is advancing, it is more and more common that such engineering solutions follow the emerging technology that can be incorporated into geotechnical engineering research and practice to develop the most efficient, most viable, and most accurate solution. In order to achieve such an elaborate task, we, as geotechnical engineers, still make use of three tools at our disposal: (i) experimental methods, (ii) numerical methods, and (iii) analytical methods.

In this introductory chapter, three emerging fields of geotechnical engineering where there have been significant developments in the last couple of decades are touched upon. Those are the (i) numerical methods in geotechnical engineering, (ii) unsaturated soils, and (iii) offshore geotechnics. A short summary of the basic notion of what each field mainly investigates and what has been recently done in each topic in recent years is also given.

Numerical methods in geotechnical engineering

Numerical methods provide approximate solutions to complex engineering problems. Soils are heterogeneous, anisotropic, and multiphase particulate materials with nonlinear stress-strain relationships. Thus, any geotechnical engineering problem containing soils as a material or as an entire domain exhibits a certain level of complexity demanding robust numerical methods to tackle the issue. Such complexity is often associated with material behavior or boundary conditions or both. As it is obvious that no analytical solutions are available to such intricate problems, numerical methods are found to be quite handy and surely the only available option that yields approximate solutions with an acceptable error margin.

Some of the commonly used numerical methods today are finite element method (FEM), finite difference method (FDM), boundary element method (BEM), discrete element method (DEM), finite volume method (FVM), material point method (MPM), hybrid methods, etc. While it is no straightforward task to decide what method to use, for most of the cases, the numerical method to employ is mostly problem-dependent. That is, the desired unknowns in the problem and the crucial details of what variables need to be calculated, where they are to be calculated and to what accuracy is wanted, will actually determine the most appropriate method.

In a brief summary of how numerical methods are employed, it should be noted that in most of these methods, the physical problem is first defined in the spatial domain, and if the problem is time-dependent, a temporal domain exists as well. Then the governing equations are written in these domains (or only along the boundaries, see BEM) in terms of the field variables which are simply the unknowns in the problem. In geotechnical engineering problems, field variables are mostly the displacements of the solid skeleton and the pore pressures in the voids of soils, that is, pore water or pore air pressures depending on whether the soil is modeled as a two-phase or a three-phase medium, respectively.

Then the governing equations are either discretized in existing domains or simply defined at specified locations in the soil material, and the necessary boundary and initial conditions are prescribed in terms of field variables (or the degrees of freedom) and their derivatives with respect to Cartesian coordinates and/or to time. When the discretized forms of the governing equations are obtained, they are to be solved using an available numerical method for each load or time step. While such a clear-cut process is sufficient to provide some sort of a numerical solution to the problem, it is important to note that the ultimate solution will only be as accurate as how much the problem is simplified in the beginning of the process.

Once the field variables are calculated, stresses (or any other stress-state variable) are computed at material points (i.e., convergence or integration points) in the soil elemental level through constitutive models using the strain field. Strains are essentially calculated utilizing the kinematic strain-displacement relations. For the soil skeleton, constitutive law is the effective stress-strain relationship, and for the soil pores, the necessary constitutive equation is the Darcy's law governing the flow of pore fluid due to differences in the total head between two points of interest in the soil.

Some of the trending problems encountered in geotechnical engineering in recent years, where numerical analysis is the key element, are:

- ⊙ Constitutive theories for transient soil response. Here, particularly the theoretical models developed for predicting the cyclic behavior of soft soils and impact-induced response of special soils and rocks entail certain issues.

- ⊙ Large deformation or post-failure behavior of soft soils. A serious issue related to soft ground is "post-liquefaction behavior" as well as settlements and lateral spreading of loosely deposited saturated sands subject to seismic motions. Another one is the pile-driving problem or modeling the penetration of piles into soil deposits. This is a typical example of a large deformation and post-failure behavior, which may pose a challenge in terms of numerical modelling.

- ⊙ Uncertainties in distributions/initial conditions of mechanical quantities of soils. Geotechnical problems always suffer from lack of knowledge on soil properties and initial conditions. Inadequate and insufficient information in geotechnical problems frequently lead to a discrepancy between prediction and measurement, which implies that the prediction in these problems is accompanied with several types of uncertainties. While deterministic approach has since been quite useful in providing an estimate solution provided that there is certain amount of data assimilation and inverse analyses back the approach, probabilistic approach is also promising. On the latter approach, stochastic analysis is a technique that considers the uncertainty as a probabilistic variable. It is employed to assess the possibility of failure or other limit states of soil-structure systems. Probabilistic approach is directly related to reliability analysis and performance-based design.

Unsaturated soils

Another emerging field in geotechnical engineering is related to unsaturated soils. The response of unsaturated soils (USS) constitutes an important consideration for many problems in geotechnical engineering and geomechanics. Current geotechnical practice that models unsaturated soil mechanical properties as if they represent fully saturated conditions results in unsafe design due to a potential rise in the water content of soil. On the other hand, saturating natural soil samples in the laboratory, which normally would represent partially saturated state with in situ water contents, softens the soil which might lead to overdesign. Since such soil layers in the field are anticipated to have varying degrees of saturation during seasonal changes, this elevates the associated risks.

The developments in the field of "unsaturated soil mechanics" have lagged behind those of "saturated soil mechanics," and the study of unsaturated soils has only recently gained importance. Although there have been many studies conducted in the last 30 years, there still needs experimental and theoretical works to be done to understand the USS hydromechanical behavior. Upon laboratory testing, predicted results should be calibrated such that the developed model could subsequently be employed to solve a practical problem through available numerical methods.

This requires the model to be implemented into a computer program in terms of numerical algorithms developed for an elemental soil using USS parametric relations in a multiparametric space. This means, the development in the USS mechanics is in direct relation with the developments in the previous topic. On the one hand, there is the objective of eliminating the downsides of treating USS as fully saturated and therefore reducing the associated unconservative design owing to unpredicted behavior of USS. On the other hand, there is the uneconomical design to be reduced due to overconservative approach.

Water content in unsaturated soils plays an important role in evaluating the hydromechanical response in relation to suction. Behaviors of USS under static loadings are affected significantly by their volume changes as a function of the change in their water content. This is, however, directly related to the difference in the pressures of pore water and pore air which is called "matric suction," s. The major stress variable controlling the deformation characteristics of USS is suction. As the matric suction increases, stiffness and shear strength of USS increase as well under constant net mean stress.

As for the stress state, while a single effective stress is sufficient to describe the stress-strain relationship of saturated soils, this is not the case for USS. Hence, effective stress equation needs to be modified in formulating a constitutive model for USS. This issue has been the main subject of debate among the researchers since the pioneering study is the first to propose a relation for effective stress that accounts for the air phase on the average stress acting on the solid skeleton. The "Bishop stress" is written as the modified effective stress:

$s" = s - cs\, E1$

where c is called the Bishop parameter which is a function of the water degree of saturation. While some researchers claim that there is a "smooth" transition to the new definition of the effective stress, others say that there needs to be two independent stress variables, that is, the net stress, s_{net} or -s , and matric suction to govern the response of USS.

It is important that in developing a constitutive model that accounts for hydromechanical behavior of USS, *hydraulic hysteresis* observed in the water retention behavior is taken into account. Such a behavior is self-evident in a typical constitutive relationship called the "soil-water characteristic curve (SWCC) or water retention curve (WRC)" irrespective of the soil type. During a drying-wetting cycle causing changes in saturation and suction, USS exhibits such hysteretic behavior. Therefore, it is necessary that the hysteretic response is incorporated in the mathematical formulation of a constitutive model for USS. Some of the common relationships are given in Table 1. There, θ is the volumetric water content, θ is the saturated water content (%), θ_r is the residual volumetric water content, and ψ is the suction head with parameters, λ, a, m, and n, being empirical constants.

Offshore geotechnics

The last field of interest that gains popularity mainly in the last two decades is offshore geotechnical engineering. Offshore structures that reside on or around seabed soil are essential for energy, protection, or transmission. The stability of the whole soil-structure system relies not only on the structural integrity under wave action but that the seabed soil must be able to withstand the induced stresses and pore pressure buildup against various loadings due to wave and currents. Geotechnical considerations are important in identifying such conditions leading to instability of seabed-offshore structure systems under wave action. Therefore, it is important that wave-induced soil behavior is modeled via appropriate theoretical frameworks.

The developments in this field depend upon three interrelated subtopics: (i) wave mechanics, (ii) structural mechanics, and (iii) soil mechanics. Hence, the fluid-soil-structure interaction (FSSI) is an indispensable field of study that one who is working on offshore geotechnics needs to tackle with diligence. In addition, the studies made in this field are also related to the topics discussed in the previous sections in that numerical methods are frequently employed to provide approximate engineering solutions to related problems. Also, while it is not common to observe unsaturated soils in offshore environments (except the air voids made up due to sea shells), aspects of soil constitutive modeling associated with unsaturated soils are indirectly applicable to the elemental-level response of saturated soils encountered in seabed soils. Thus, in this section of this introductory chapter, the focus is on such relationships between seabed soils and their constitutive behavior under cyclic wave loading as opposed to specific offshore structures and their analysis details. Interested readers can refer to the recent proceedings published by the International Society of Offshore and Polar Engineers (ISOPE).

INTRODUCTION TO SOIL MECHANICS

Soil mechanics is one of the engineering disciplines that deals with soils as and engineering material. Since engineering ages, engineers have been handling soils as an engineering material for various construction projects. During 18th and 19th centuries, some modern engineering theories have been employed in this field, following the development of Newtonian mechanics, Coulomb's and Rankine's lateral earth pressure theories.

Related terminologies of soil mechanics are foundation engineering, geotechnical engineering, and geo environmental engineering. Foundation engineering is the field of designing safe foundations, including building footings and retaining structures, and the construction of earth structures such as embankments, earth or rock fill dams, and safe earth slopes etc., based on the knowledge of soil mechanics. Thus, the discipline has been called soil mechanics and foundation engineering for many years.

The new term geotechnical engineering was coined around 1970 to merge rock mechanics into soil mechanics and foundation engineering, and is the most popularly used terminology in this field at present. In the 1980's environmental related geotechnical engineering became a great engineering concern, and the term geo environmental engineering was created, which includes the design and construction of solid and liquid waste containment facilities, and any other environmentally related geotechnical engineering problems.

FATHER OF SOIL MECHANICS

Dr. Karl Von Terzaghi was born in Prague, Austria in 1883. The modern era of soil mechanics had to wait until Dr. Terzaghi published a book called *Erdbaumechanik*. Especially, his new concept of "effective stress", which deals with interaction with pore water, has revolutionized the mechanics of soils. The development of modern soil mechanics is due to his great contribution. He is now regarded as the father of soil mechanics.

UNIQUENESS OF SOILS

Soil can be defined as assemblage of nonmetallic solid particles (mineral grains), and it consists of three phases: solid, liquid (water), and gas (air). Commonly used terms such as gravel, sand, silt, and clay are the names of soils based on their particle grain sizes. Quartz, mica, feldspar, etc., are based on their crystal names. Soil is a very unique material and complex in nature. The unique characteristics of soils are as follows:

1. It is not a solid continuous material, but is rather composed of three different constituents
2. Particle sizes have significant influence on soil behavior from granular soil to clay.
3. The amount of water also plays a very important role in soil behavior.
4. Its stress-strain relation is not linear from the small strain levels.
5. Its pore spaces possess the capability of water flow.
6. It has time-dependent characteristics.
7. It swells when wetted or shrinks when dried.
8. It is an anisotropic material due to plastic shapes
9. It is also spatially non homogeneous.

ANALYSIS APPROACHES

Complexity and spatial variation of soil makes the field observation and laboratory testing very significant.

Field observation ranges from geological study of the site to soil sampling and sometimes in situ testing of properties. Sampled specimens are brought back to laboratories for physical and mechanical tests. Based on field observations and laboratory test data, geotechnical engineers classify soils; determine design properties and design safe foundations and earth structures, by fully utilizing modern soil mechanics knowledge and foundation engineering concepts. Afterward, construction companies carry out construction of the project according to the specifications. The last stage is field monitoring of performance of earth structures.

Soil is a natural body consisting of layers (soil horizons) that are primarily composed of minerals which differ from their parent materials in their texture, structure, consistency, color, chemical, biological and other characteristics. It is the unconsolidated or loose covering of fine rock particles that covers the surface of the earth. Soil is the end product of the influence of the climate (temperature, precipitation), relief (slope), organisms (flora and fauna), parent materials (original minerals), and time. In engineering terms, soil is referred to as regolith, or loose rock material that lies above the 'solid geology'. In horticulture, the terms 'soil' is defined as the layer that contains organic material that influences and has been influenced by plant roots and may range in depth from centimetres to many metres.

Soil is composed of particles of broken rock (parent materials) which have been altered by physical, chemical and biological processes that include weathering (disintegration) with associated erosion (movement). Soil is altered from its parent material by the interactions between the lithosphere, hydrosphere, atmosphere, and biosphere. It is a mixture of mineral and organic materials in the form of solids, gases and liquids. Soil is commonly referred to as "earth" or "dirt"; technically, the term "dirt" should be restricted to displaced soil.

Soil forms a structure filled with pore spaces and can be thought of as a mixture of solids, water, and gases. Accordingly, soils are often treated as a three-state system. Most soils have a density between 1 and 2 g/cm^3. Little of the soil of planet Earth is older than the Pleistocene and none is older than the Cenozoic, although fossilised soils are preserved from as far back as the Archean.

DEFINITION OF SOIL & APPROACHES OF SOIL STUDY

Soil is a natural body of mineral and organic constituents differentiated into horizons usually unconsolidated, of variable depth which differs among themselves as well as from the underlying parent material in morphology, physical makeup, chemical properties and composition and biological characteristics.

The unconsolidated mineral matter on the surface of the earth that has been subjected to and influenced by genetic and environmental factors of parent material, climate (including moisture and temperature effects), macro and microorganisms and topography, all affecting over a period of time and producing a product, that is "SOIL" that differs from the material from which it is derived in many, physical, chemical, biological and morphological properties and characteristics.

The unconsolidated mineral material on the immediate surface of the earth that serves as a natural medium for the growth of land plants.

APPROACHES OF SOIL STUDY

Two Concepts: One treats soil as a natural body, weathered and synthesized product in nature while other treats soil as a medium for plant growth.

1. **Pedological Approach:** The origin of the soil, its classification and its description are examined in Pedology. (From Greek word pedon, means soil or earth). Pedology is the study of soil as a natural body and does not focus on the soil's immediate practical use. A pedologist studies, examines and classifies soil as they occur in their natural environment.

2. **Edaphological Approach:** Edophology (from Greek word edaphos, means soil or ground) is the study of soil from the stand point of higher plants. Edaphologists consider the various properties of soil in relation to plant production. They are practical and have the production of food and fiber as their ultimate goal. They must determine the reasons for variation in the productivity of soils and find means for improvement.

STUDIES CONCERNING SOIL FERTILITY

The history of the study of soil is intimately tied to our urgent need to provide food for ourselves and forage for our animals. Throughout history, civilizations have prospered or declined as a function of the availability and productivity of their soils.

The Greek historian Xenophon (450–355 B.C.) is credited with being the first to expound upon the merits of green-manuring crops: "But then whatever weeds are upon the ground, being turned into earth, enrich the soil as much as dung."

Columella's "Husbandry," circa 60 A.D., was used by 15 generations (450 years) under the Roman Empire until its collapse. From the fall of Rome to the French Revolution, knowledge of soil and agriculture was passed on from parent to child and as a result, crop yields were low. During the European Dark Ages, Yahya Ibn_al-'Awwam's handbook guided the people of North Africa, Spain and the Middle East with its emphasis on irrigation; a translation of this work was finally carried to the southwest of the United States.

Experiments into what made plants grow first led to the idea that the ash left behind when plant matter was burned was the essential element but overlooked the role of nitrogen, which is not left on the ground after combustion. In about 1635, the Flemish chemist Jan Baptist van Helmont thought he had proved water to be the essential element from his famous five years' experiment with a willow tree grown with only the addition of rainwater. His conclusion came from the fact that the increase in the plant's weight had been produced only by the addition of water, with no reduction in the soil's weight. John Woodward (d. 1728) experimented with various types of water ranging from clean to muddy and found muddy water the best, and so he concluded that earthy matter was the essential element. Others concluded it was humus in the soil that passed some essence to the growing plant. Others held that the vital growth principal was something passed from dead plants or animals to the new plants. At the start of the 18th century, Jethro Tull demonstrated that it was beneficial to cultivate the soil, but his opinion that the stirring made the fine parts of soil available for plant absorption was erroneous.

As chemistry developed, it was applied to the investigation of soil fertility. The French chemist Antoine Lavoisier showed in about 1778 that plants and animals must "combust" oxygen internally to live and was able to deduce that most of the 165-pound weight of van Helmont's willow tree derived from air. It was the French agriculturalist Jean-Baptiste Boussingault who by means of experimentation obtained evidence showing that the main sources of carbon, hydrogen and oxygen for plants were the air and water. Justus von Liebig in his book *Organic Chemistry in its Applications to Agriculture and Physiology* (published 1840), he asserted that the chemicals in plants must have come from the soil and air and that to maintain soil fertility, the used minerals must be replaced. Liebig nevertheless believed the nitrogen was supplied from the air. The enrichment of soil with guano by the Incas was rediscovered in 1802, by Alexander von Humboldt. This led to its mining and that of Chilean nitrate and to its application to soil in the United States and Europe after 1840.

The work of Liebig was a revolution for agriculture, and so other investigators started experimentation based on it. In England John Bennet Lawes and Joseph Henry Gilbert worked in the Rothamsted Experimental Station, founded by the former, and discovered that plants took nitrogen from the soil, and that salts needed to be in an available state to be absorbed by plants. Their investigations also produced the "superphosphate", consisting in the acid treatment of phosphate rock. This led to the invention and use of salts of potassium (K) and nitrogen (N) as fertilizers. Finally, the chemical basis of nutrients delivered to the soil in manure was understood and in the mid-19th century chemical fertilisers were applied. The dynamic interaction of soil and its life forms awaited discovery.

In 1856 J. T. Way discovered that ammonia contained in fertilisers was transformed into nitrates, and twenty years later R. W. Warington proved that this transformation was done by living organisms. In 1890 Sergei Winogradsky announced he had found the bacteria responsible for this transformation.

It was known that certain legumes could take up nitrogen from the air and fix it to the soil. The development of bacteriology towards the end of the 19th century led to an understanding of the role played in nitrogen fixation by bacteria. The symbiosis of bacteria and leguminous roots, and the fixation of nitrogen by the bacteria, were simultaneously discovered by German agronomist Hermann Hellriegel and Dutch microbiologist Martinus Beijerinck.

Crop rotation, mechanisation, chemical and natural fertilisers led to a doubling of wheat yields in Western Europe between 1800 and 1900.

STUDIES CONCERNING SOIL FORMATION

The scientists who studied the soil in connection with agricultural practices had considered it mainly as a static substrate. However, soil is the result of evolution from more ancient geological materials. Other scientists later began to study soil genesis and as a result also soil types and classifications. In 1860, in Mississippi, Eugene W. Hilgard studied the relationship among rock material, climate, and vegetation, and the type of soils that were developed. He realised that the soils were dynamic, and considered soil types classification.

Unfortunately his work was not continued. At the same time Vasily Dokuchaev was leading a team of soil scientists in Russia who conducted an extensive survey of soils, finding that similar basic rocks, climate and vegetation types lead to similar soil layering and types, and established the concepts for soil classifications. Curtis F. Marbut was influenced by the work of the Russian team, translated Glinka's publication into English, and as he was placed in charge of the U. S. National Cooperative Soil Survey, applied it to a national soil classification system.

PARENT MATERIAL AND SOIL FORMATION

Soil formation, or pedogenesis, is the combined effect of physical, chemical, biological and anthropogenic processes on soil parent material. Soil is said to be formed when organic matter has accumulated and colloids are washed downward, leaving deposits of clay, humus, iron oxide, carbonate, and gypsum. As a result, horizons form in the soil profile. These constituents are moved (translocated) from one level to another by water and animal activity. The alteration and movement of materials within a soil causes the formation of distinctive soil horizons.

How soil formation proceeds is influenced by at least five classic factors that are intertwined in the evolution of a soil. They are: **parent material**, **climate**, **topography** (relief), **organisms**, and **time**. When reordered to climate, relief, organisms, parent material, and time, they form the acronym CROPT.

An example of the development of a soil would begin with the weathering of lava flow bedrock, which would produce the purely mineral-based parent material from which the soil texture forms. Soil development would proceed most rapidly from bare rock of recent flows in a warm climate, under heavy and frequent rainfall. Under such conditions, plants become established very quickly on basaltic lava, even though there is very little organic material. The plants are supported by the porous rock as it is filled with nutrient-bearing water that carries dissolved minerals from the rocks and guano. Crevasses and pockets, local topography of the rocks, would hold fine materials and harbour plant roots. The developing plant roots are associated with mycorrhizal fungi that assist in breaking up the porous lava, and by these means organic matter and a finer mineral soil accumulate with time.

PARENT MATERIAL

The mineral material from which a soil forms is called parent material. Rock, whether its origin is igneous, sedimentary, or metamorphic, is the source of all soil mineral materials and origin of all plant nutrients with the exceptions of nitrogen, hydrogen and carbon. As the parent material is chemically and physically **weathered**, **transported**, deposited and precipitated, it is transformed into a soil.

Typical soil mineral materials are:

- Quartz: SiO_2
- Calcite: $CaCO_3$
- Feldspar: $KAlSi_3O_8$
- Mica (biotite): $K(Mg,Fe)_3AlSi_3O_{10}(OH)_2$

CLASSIFICATION OF PARENT MATERIAL

Parent materials are classified according to how they came to be deposited. Residual materials are mineral materials that have weathered in place from primary bedrock. Transported materials are those that have been deposited by water, wind, ice or gravity. And cumulose material is organic matter that has grown and accumulates in place.

Residual soils are soils that develop from their underlying parent rocks and have the same general chemistry as those rocks. The soils found on mesas, plateaux, and plains are residual soils. In the United States as little as three percent of the soils are residual.

Most soils derive from **transported** materials that have been moved many miles by wind, water, ice and gravity.

- Aeolian processes (movement by wind) are capable of moving silt and fine sand many hundreds of miles, forming loess soils (60–90 percent silt), common in the Midwest of North America and in Central Asia. Clay is seldom moved by wind as it forms stable aggregates.

- Water-transported materials are classed as either alluvial, lacustrine, or marine. Alluvial materials are those moved and deposited by flowing water. Sedimentary deposits settled in lakes are called lacustrine. Lake Bonneville and many soils around the Great Lakes of the United States are examples. Marine deposits, such as soils along the Atlantic and Gulf Coasts and in the Imperial Valley of California of the United States, are the beds of ancient seas that have been revealed as the land uplifted.

⊙ Ice moves parent material and makes deposits in the form of terminal and lateral moraines in the case of stationary glaciers. Retreating glaciers leave smoother ground moraines and in all cases, outwash plains are left as alluvial deposits are moved downstream from the glacier.

⊙ Parent material moved by gravity is obvious at the base of steep slopes as talus cones and is called colluvial material.

Cumulose parent material is not moved but originates from deposited organic material. This includes peat and muck soils and results from preservation of plant residues by the low oxygen content of a high water table. While peat may form sterile soils, muck soils may be very fertile.

Weathering of parent material

The weathering of parent material takes the form of **physical disintegrating** and **chemical decomposition** and transformation.

⊙ Physical disintegration is the first stage in the transformation of parent material into soil. The freezing of absorbed water causes the physical splitting of material along a path toward the center of the rock, while temperature gradients within the rock can cause exfoliation of "shells". Cycles of wetting and drying cause soil particles to be ground down to a finer size, as does the physical rubbing of material caused by wind, water, and gravity. Organisms also reduce parent material in size through the action of plant roots or digging on the part of animals.

⊙ Chemical decomposition results when minerals are made soluble by water or are changed in structure. The first three of the following list are solubility changes and the last three are structural changes:

1. The solution of salts in water results from the action of bipolar water on ionic salt compounds.
2. Hydrolysis is the transformation of minerals into polar molecules by the splitting of the mineral and the intervening water. This results in soluble acid-base pairs. For example, the hydrolysis of orthoclase-feldspar transforms it to acid silicate clay and basic potassium hydroxide, which are more soluble.
3. In carbonation, the reaction of carbon dioxide in solution with water forms carbonic acid. Carbonic acid will transform calcite into more soluble calcium bicarbonate.
4. Hydration is the inclusion of water in a mineral structure, causing it to swell and leaving it more stressed and easily decomposed.
5. Oxidation of a mineral compound causes it to swell and increase its oxidation number, leaving it more easily attacked by water or carbonic acid.

6. Reduction means the oxidation number of some part of the mineral is reduced, which occurs when oxygen is scarce. The reduction of minerals leaves them electrically unstable, more soluble and internally stressed and easily decomposed.

Of the above, hydrolysis and carbonation are the most effective.

Saprolite is a particular example of a residual soil formed from the transformation of granite, metamorphic and other types of bedrock into clay minerals. Often called "weathered granite", saprolite is the result of weathering processes that include: hydrolysis, chelation from organic compounds, hydration (the solution of minerals in water with resulting cation and anion pairs) and physical processes that include freezing and thawing. The mineralogical and chemical composition of the primary bedrock material, its physical features, including grain size and degree of consolidation, and the rate and type of weathering transform the parent material into a different mineral. The texture, pH and mineral constituents of saprolite are inherited from its parent material.

CLIMATE

Climate is the dominant factor in soil formation, and soils show the distinctive characteristics of the climate zones in which they form. Mineral precipitation and temperature are the primary climatic influences on soil formation.

The direct influences of climate include:

- A shallow accumulation of lime in low rainfall areas as caliche
- Formation of acid soils in humid areas
- Erosion of soils on steep hillsides
- Deposition of eroded materials downstream
- Very intense chemical weathering, leaching, and erosion in warm and humid regions where soil does not freeze

Climate directly affects the rate of weathering and leaching. Soil is said to be formed when detectable layers of clays, organic colloids, carbonates, or soluble salts have been moved downward. Wind moves sand and smaller particles, especially in arid regions where there is little plant cover. The type and amount of precipitation influence soil formation by affecting the movement of ions and particles through the soil, and aid in the development of different soil profiles. Soil profiles are more distinct in wet and cool climates, where organic materials may accumulate, than in wet and warm climates, where organic materials are rapidly consumed. The effectiveness of water in weathering parent rock material depends on seasonal and daily temperature fluctuations. Cycles of freezing and thawing constitute an effective mechanism which breaks up rocks and other consolidated materials.

Climate also indirectly influences soil formation through the effects of vegetation cover and biological activity, which modify the rates of chemical reactions in the soil.

TOPOGRAPHY

The topography, or relief, characterised by the inclination of the surface, determines the rate of precipitation runoff and rate of formation or erosion of the surface soil profiles. Steep slopes allow rapid runoff and erosion of the top soil profiles and little mineral deposition in lower profiles. Depressions allow the accumulation of water, minerals and organic matter and in the extreme, the resulting soils will be saline marshes or peat bogs. Intermediate topography affords the best conditions for the formation of an agriculturally productive soil.

ORGANISMS

Soil is the most abundant ecosystem on Earth, but the vast majority of organisms in soil are microbes, a great many of which have not been described. There may be a population limit of around one billion cells per gram of soil, but estimates of the number of species vary widely. One estimate put the number at over a million species per gram of soil, although a later study suggests a maximum of just over 50,000 species per gram of soil. The total number of organisms and species can vary widely according to soil type, location, and depth.

Plants, animals, fungi, bacteria and humans affect soil formation. Animals, soil mesofauna and micro-organisms mix soils as they form burrows and pores, allowing moisture and gases to move about. In the same way, plant roots open channels in soils. Plants with deep taproots can penetrate many metres through the different soil layers to bring up nutrients from deeper in the profile. Plants with fibrous roots that spread out near the soil surface have roots that are easily decomposed, adding organic matter. Micro-organisms, including fungi and bacteria, effect chemical exchanges between roots and soil and act as a reserve of nutrients. Humans impact soil formation by removing vegetation cover with erosion as the result. Their tillage also mixes the different soil layers, restarting the soil formation process as less weathered material is mixed with the more developed upper layers.

Vegetation impacts soils in numerous ways. It can prevent erosion caused by excessive rain that results in surface runoff. Plants shade soils, keeping them cooler and slowing evaporation of soil moisture, or conversely, by way of transpiration, plants can cause soils to lose moisture. Plants can form new chemicals that can break down minerals and improve soil structure. The type and amount of vegetation depends on climate, topography, soil characteristics, and biological factors. Soil factors such as density, depth, chemistry, pH, temperature and moisture greatly affect the type of plants that can grow in a given location. Dead plants and fallen leaves and stems begin their decomposition on the surface. There, organisms feed on them and mix the organic material with the upper soil layers; these added organic compounds become part of the soil formation process.

TIME

Time is a factor in the interactions of all the above. Over time, soils evolve features that are dependent on the interplay of other soil forming factors. Soil is always changing. It takes about 800 to 1000 years for a 2.5 cm (0.98 in) thick layer of fertile soil to be formed in nature. For example, recently deposited material from a flood exhibits no soil development because there has not been enough time for the material to form a structure that further defines soil. The original soil surface is buried, and the formation process must begin anew for this deposit. Over a period of between hundreds and thousands of years, the soil will develop a profile that depends on the intensities of biota and climate. While soil can achieve relative stability of its properties for extended periods, the soil life cycle ultimately ends in soil conditions that leave it vulnerable to erosion. Despite the inevitability of soil retrogression and degradation, most soil cycles are long.

Soil-forming factors continue to affect soils during their existence, even on "stable" landscapes that are long-enduring, some for millions of years. Materials are deposited on top or are blown or washed from the surface. With additions, removals and alterations, soils are always subject to new conditions. Whether these are slow or rapid changes depends on climate, topography and biological activity.

PHYSICAL PROPERTIES OF SOILS

The physical properties of soils, in order of decreasing importance, are texture, structure, density, porosity, consistency, temperature, colour and resistivity. Most of these determine the aeration of the soil and the ability of water to infiltrate and to be held in the soil. Soil texture is determined by the relative proportion of the three kinds of soil particles, called soil "separates": sand, silt, and clay. Larger soil structures called "peds" are created from the separates when iron oxides, carbonates, clay, and silica with the organic constituent humus, coat particles and cause them to adhere into larger, relatively stable secondary structures. Soil density, particularly bulk density, is a measure of soil compaction. Soil porosity consists of the part of the soil volume occupied by air and water. Consistency is the ability of soil to stick together. Soil temperature and colour are self-defining. Resistivity refers to the resistance to conduction of electric currents and affects the rate of corrosion of metal and concrete structures. The properties may vary through the depth of a soil profile.

Texture

The mineral components of soil, sand, silt and clay, determine a soil's texture. In the illustrated USDA textural classification triangle, the only soil that does not exhibit one of these predominately is called "loam". While even pure sand, silt or clay may be considered a soil, from the perspective of food production a loam soil with a small amount of organic material is considered ideal. The mineral constituents of a loam soil might be 40% sand, 40% silt and the balance 20% clay by weight. Soil texture affects soil behaviour, in particular its retention capacity for nutrients and water.

Sand and silt are the products of physical and chemical weathering; clay, on the other hand, is a product of chemical weathering but often forms as a secondary mineral precipitated from dissolved minerals. It is the specific surface area of soil particles and the unbalanced ionic charges within them that determine their role in the cation exchange capacity of soil, and hence its fertility. Sand is least active, followed by silt; clay is the most active. Sand's greatest benefit to soil is that it resists compaction and increases porosity. Silt is mineralogically like sand but with its higher specific surface area it is more chemically active than sand. But it is the clay content, with its very high specific surface area and generally large number of negative charges, that gives a soil its high retention capacity for water and nutrients. Clay soils also resist wind and water erosion better than silty and sandy soils, as the particles are bonded to each other.

Sand is the most stable of the mineral components of soil; it consists of rock fragments, primarily quartz particles, ranging in size from 2.0 to 0.05 mm (0.079 to 0.0020 in) in diameter. Silt ranges in size from 0.05 to 0.002 mm (0.002 to 0.00008 in). Clay cannot be resolved by optical microscopes as its particles are 0.002 mm (7.9×10^{-5} in) or less in diameter. In medium-textured soils, clay is often washed downward through the soil profile and accumulates in the subsoil.

Soil components larger than 2.0 mm (0.079 in) are classed as rock and gravel and are removed before determining the percentages of the remaining components and the texture class of the soil, but are included in the name. For example, a sandy loam soil with 20% gravel would be called gravelly sandy loam.

When the organic component of a soil is substantial, the soil is called organic soil rather than mineral soil. A soil is called organic if:

1. Mineral fraction is 0% clay and organic matter is 20% or more
2. Mineral fraction is 0% to 50% clay and organic matter is between 20% and 30%
3. Mineral fraction is 50% or more clay and organic matter 30% or more.

Structure

The clumping of the soil textural components of sand, silt and clay forms **aggregates** and the further association of those aggregates into larger units forms soil structures called **peds**. The adhesion of the soil textural components by organic substances, iron oxides, carbonates, clays, and silica, and the breakage of those aggregates due to expansion-contraction, freezing-thawing, and wetting-drying cycles, shape soil into distinct geometric forms.

These peds evolve into units which may have various shapes, sizes and degrees of development. A soil clod, however, is not a ped but rather a mass of soil that results from mechanical disturbance. The soil structure affects aeration, water movement, conduction of heat, plant root growth and resistance to erosion. Water has the strongest effect on soil structure due to its solution and precipitation of minerals and its effect on plant growth.

Soil structure often gives clues to its texture, organic matter content, biological activity, past soil evolution, human use, and the chemical and mineralogical conditions under which the soil formed. While texture is defined by the mineral component of a soil and is an innate property of the soil that does not change with agricultural activities, soil structure can be improved or destroyed by the choice and timing of farming practices.

Colour

Soil colour is often the first impression one has when viewing soil. Striking colours and contrasting patterns are especially noticeable. The Red River (Mississippi watershed) carries sediment eroded from extensive reddish soils like Port Silt Loam in Oklahoma. The Yellow River in China carries yellow sediment from eroding loess soils. Mollisols in the Great Plains of North America are darkened and enriched by organic matter. Podsols in boreal forests have highly contrasting layers due to acidity and leaching.

In general, colour is determined by organic matter content, drainage conditions, and the degree of oxidation. Soil colour, while easily discerned, has little use in predicting soil characteristics. It is of use in distinguishing boundaries within a soil profile, determining the origin of a soil's parent material, as an indication of wetness and waterlogged conditions, and as a qualitative means of measuring organic, salt and carbonate contents of soils. Colour is recorded in the Munsell color system as for instance 10YR3/4.

Soil colour is primarily influenced by soil mineralogy. Many soil colours are due to various **iron** minerals. The development and distribution of colour in a soil profile result from chemical and biological weathering, especially redox reactions. As the primary minerals in soil parent material weather, the elements combine into new and colourful compounds. Iron forms secondary minerals of a yellow or red colour, organic matter decomposes into black and brown compounds, and manganese, sulfur and nitrogen can form black mineral deposits.

These pigments can produce various colour patterns within a soil. Aerobic conditions produce uniform or gradual colour changes, while reducing environments (anaerobic) result in rapid colour flow with complex, mottled patterns and points of colour concentration.

Resistivity

Soil resistivity is a measure of a soil's ability to retard the conduction of an electric current. The electrical resistivity of soil can affect the rate of galvanic corrosion of metallic structures in contact with the soil. Higher moisture content or increased electrolyte concentration can lower resistivity and increase conductivity, thereby increasing the rate of corrosion. Soil resistivity values typically range from about 2 to 1000 $\Omega\bullet$m, but more extreme values are not unusual.

SOIL WATER

Water affects soil formation, structure, stability and erosion but is of primary concern with respect to plant growth. Water is essential to plants for four reasons:

1. It constitutes 85%-95% of the plant's protoplasm.
2. It is essential for photosynthesis.
3. It is the solvent in which nutrients are carried to, into and throughout the plant.
4. It provides the turgidity by which the plant keeps itself in proper position.

In addition, water alters the soil profile by dissolving and redepositing minerals, often at lower levels, and possibly leaving the soil sterile in the case of extreme rainfall and drainage.

In a loam soil, solids constitute half the volume, air one-quarter of the volume, and water one-quarter of the volume, of which only half will be available to most plants.

Water retention forces

Water is retained in a soil when the **adhesive** force of attraction of water for soil particles and the **cohesive** forces water feels for itself are capable of resisting the force of gravity which tends to drain water from the soil. When a field is flooded, the air space is displaced by water. The field will drain under the force of gravity until it reaches what is called **field capacity**, at which point the smallest pores are filled with water and the largest with water and air. The total amount of water held when field capacity is reached is a function of the specific surface area of the soil particles. As a result, high clay and high organic soils have higher field capacities. The total force required to pull or push water out of soil is termed **suction** and usually expressed in units of bars (10^5 pascal) which is just a little less than one-atmosphere pressure. Alternatively, the terms "tension" or "moisture potential" may be used.

Moisture Classification

The forces with which water is held in soils determine its availability to plants. Forces of adhesion hold water strongly to mineral and humus surfaces and less strongly to itself by cohesive forces. A plant's root may penetrate a very small volume of water that is adhering to soil and be initially able to draw in water that is only lightly held by the cohesive forces. But as the droplet is drawn down, the forces of adhesion of the water for the soil particles make reducing the volume of water increasingly difficult until the plant cannot produce sufficient suction to use the remaining water. The remaining water is considered unavailable. The amount of available water depends upon the soil texture and humus amounts and the type of plant attempting to use the water. Cacti, for example, can produce greater suction than can agricultural crop plants.

Soil Organism

Soil organism, any organism inhabiting the soil during part or all of its life. Soil organisms, which range in size from microscopic cells that digest decaying organic material to small mammals that live primarily on other soil organisms, play an important role in maintaining fertility, structure, drainage, and aeration of soil. They also break down plant and animal tissues, releasing stored nutrients and converting them into forms usable by plants. Some soil organisms are pests. Among the soil organisms that are pests of crops are nematodes, slugs and snails, symphylids, beetle larvae, fly larvae, caterpillars, and root aphids. Some soil organisms cause rots, some release substances that inhibit plant growth, and others are hosts for organisms that cause animal diseases.

Since most of the functions of soil organisms are beneficial, earth with large numbers of organisms in it tends to be fertile; one square metre of rich soil can harbour as many as 1,000,000,000 organisms.

Soil organisms are commonly divided into five arbitrary groups according to size, the smallest of which are the protists—including bacteria, actinomycetes, and algae. Next are the microfauna, which are less than 100 microns in length and generally feed upon other microorganisms. The microfauna include single-celled protozoans, some smaller flatworms, nematodes, rotifers, and tardigrades (eight-legged invertebrates). The mesofauna are somewhat larger and are heterogeneous, including creatures that feed on microorganisms, decaying matter, and living plants. The category includes nematodes, mites, springtails (wingless insects so called for the springing organ which enables them to leap), the insectlike proturans, which feed on fungi, and the pauropods.

The fourth group, the macrofauna, are also quite diverse. The most common example is the potworm, a white, segmented worm that feeds on fungi, bacteria, and decaying plant material. The group also includes slugs, snails, and millipedes, which feed on plants, and centipedes, beetles and their larvae, and the larvae of flies, which feed on other organisms or on decaying matter.

Megafauna constitute the largest soil organisms and include the largest earthworms, perhaps the most important creatures that live in the topsoil. Earthworms pass both soil and organic matter through their guts, in the process aerating the soil, breaking up the litter of organic material on its surface, and moving material vertically from the surface to the subsoil. This is extremely important to soil fertility, and it develops the structure of the soil as a matrix for plants and other organisms. It has been estimated that earthworms completely turn over the equivalent of all the soil on the planet to a depth one inch (2.5 cm) every 10 years. Some vertebrates are also in the megafauna category; these include all sorts of burrowing animals, such as snakes, lizards, gophers, badgers, rabbits, hares, mice, and moles.

One of the most important roles of soil organisms is breaking up the complex substances in decaying plants and animals so that they can be used again by living plants. This involves soil organisms as catalysts in a number of natural cycles, among the most prominent being the carbon, nitrogen, and sulfur cycles.

The carbon cycle begins in plants, which combine carbon dioxide from the atmosphere with water to make plant tissues such as leaves, stems, and fruits. Animals eat the plants and convert the tissues into animal tissues. The cycle is completed when the animals die and their decaying tissues are eaten by soil organisms, a process that releases carbon dioxide.

Proteins are the basic stuff of organic tissues, and nitrogen is an essential element of all proteins. The availability of nitrogen in forms that plants can use is a basic determinant of the fertility of soils; the role of soil organisms in facilitating the nitrogen cycle is therefore of great importance. When a plant or animal dies, soil organisms break up the complex proteins, polypeptides, and nucleic acids in their bodies and produce ammonium, ions, nitrates, and nitrites that plants then use to build their body tissues.

Both bacteria and blue-green algae can fix nitrogen directly from the atmosphere, but this is less vital to plant development than the symbiotic relationship between the bacteria genus *Rhizobium* and leguminous plants and certain trees and shrubs. In return for secretions from their host that encourage their growth and multiplication, *Rhizobia* fix nitrogen in nodules of the host plant's roots, providing nitrogen in a form usable by the plant.

Soil organisms also participate in the sulfur cycle, mostly by breaking up the naturally abundant sulfur compounds in the soil so that this vital element is available to plants. The smell of rotten eggs so common in swamps and marshes is due to the hydrogen sulfide produced by these microorganisms.

Though soil organisms have become less important in agriculture due to the development of synthetic fertilizers, they play a vital role in woodlands, especially in the creation of humus, a finely separated complex of organic materials composed of decaying leaves and other vegetable matter.

When a leaf falls it cannot be eaten by most animals. After the water-soluble components of the leaf are leached out, fungi and other microflora attack its structure, making it soft and pliable. Now the litter is palatable to a wide variety of invertebrates, which fragment it into a mulch. The multipedes, wood lice, fly larvae, springtails, and earthworms leave the litter relatively unchanged organically, but they create a suitable substrate for the growth of the primary decomposers that break it into simpler chemical compounds. There is also a group called secondary decomposers (some creatures, such as the springtails, are in both groups), which break it down even further.

So the organic matter of leaves is constantly being digested and redigested by waves of increasingly smaller organisms. Eventually the humic substance that remains may be as little as one-fourth of the original organic matter of the litter. Gradually this humus is mixed into the soil by burrowing animals (such as moles, rabbits, etc.) and by the action of the earthworms.

next most prevalent. Earthworms, millipedes, centipedes, and insects make up most of the rest of the larger soil animal species. Plant roots also make a significant contribution to the biomass—the combined root length from a single plant can exceed 600 km (373 miles) in the top metre of a soil profile.

The soil flora and fauna play an important role in soil development. Microbiological activity in the rooting zone of soils is important to soil acidity and to the cycling of nutrients. Aerobic and anaerobic (oxygen-depleted) microniches support microbes that determine the rate of the production of carbon dioxide (CO_2) from organic matter or of nitrate (NO_3^-) from molecular nitrogen (N_2).

The carbon and nitrogen cycles are two important microbe-mediated cycles that are described in more detail in the section Soils in ecosystems. In this section, however, it is worth pointing out how they illustrate the complex, integrated nature of a soil's physical, chemical, and biological behaviour: soil peds and pore spaces provide microniches for the action of carbon- and nitrogen-cycling organisms, soil humus provides the nutrient reservoirs, and soil biomass provides the chemical pathways for cycling. The carbon in dead biomass is converted to CO_2 by aerobic microorganisms and to organic acids or alcohols by anaerobic microorganisms. Under highly anaerobic conditions, methane (CH_4) is produced by bacteria. The CO_2 produced can be used by photosynthetic microorganisms or by higher plants to create new biomass and thus initiate the carbon cycle again.

The nitrogen (N) bound into proteins in dead biomass is consumed by microorganisms and converted into ammonium ions (NH_4^+) that can be directly absorbed by plant roots (for example, lowland rice). The ammonium ions are usually converted to nitrite ions (NO_2^-) by *Nitrosomonas* bacteria, followed by a second conversion to nitrate (NO_3^-) by *Nitrobacter* bacteria. This very mobile form of nitrogen is that most commonly absorbed by plant roots, as well as by microorganisms in soil. To close the nitrogen cycle, nitrogen gas in the atmosphere is converted to biomass nitrogen by *Rhizobium* bacteria living in the root tissues of legumes (e.g., alfalfa, peas, and beans) and leguminous trees (such as alder) and by cyanobacteria and *Azotobacter* bacteria.

SOIL FORMATION

As stated at the beginning of this article, soils evolve under the action of biological, climatic, geologic, and topographic influences. The evolution of soils and their properties is called soil formation, and pedologists have identified five fundamental soil formation processes that influence soil properties. These five "state factors" are parent material, topography, climate, organisms, and time.

Parent Material

Parent material is the initial state of the solid matter making up a soil. It can consist of consolidated rocks, and it can also include unconsolidated deposits such as river alluvium, lake or marine sediments, glacial tills, loess (silt-sized, wind-deposited particles), volcanic ash, and organic matter (such as accumulations in swamps or bogs). Parent materials influence soil formation through their mineralogical composition, their texture, and their stratification (occurrence in layers). Dark-coloured ferromagnesian (iron- and magnesium-containing) rocks, for example, can produce soils with a high content of iron compounds and of clay minerals in the kaolin or smectite groups, whereas light-coloured siliceous (silica-containing) rocks tend to produce soils that are low in iron compounds and that contain clay minerals in the illite or vermiculite groups. The coarse texture of granitic rocks leads to a coarse, loamy soil texture and promotes the development of E horizons (the leached lower regions of the topmost soil layer). The fine texture of basaltic rocks, on the other hand, yields soils with a loam or clay-loam texture and hinders the development of E horizons. Because water percolates to greater depths and drains more easily through soils with coarse texture, clearly defined E horizons tend to develop more fully on coarse parent material.

In theory, parent material is either freshly exposed solid matter (for example, volcanic ash immediately after ejection) or deep-lying geologic material that is isolated from atmospheric water and organisms. In practice, parent materials can be deposited continually by wind, water, or volcanoes and can be altered from their initial, isolated state, thereby making identification difficult. If a single parent material can be established for an entire soil profile, the soil is termed monogenetic; otherwise, it is polygenetic. An example of polygenetic soils are soils that form on sedimentary rocks or unconsolidated water- or wind-deposited materials. These so-called stratified parent materials can yield soils with intermixed geologic layering and soil horizons—as occurs in southeastern England, where soils forming atop chalk bedrock layers are themselves overlain by soil layers formed on both loess and clay materials that have been modified by dissolution of the chalk below.

Adjacent soils frequently exhibit different profile characteristics because of differing parent materials. These differing soil areas are called lithosequences, and they fall into two general types. Continuous lithosequences have parent materials whose properties vary gradually along a transect, the prototypical example being soils formed on loess deposits at increasing distances downwind from their alluvial source. Areas of such deposits in the central United States or China show systematic decreases in particle size and rate of deposition with increasing distance from the source. As a result, they also show increases in clay content and in the extent of profile development from weathering of the loess particles.

By contrast, discontinuous lithosequences arise from abrupt changes in parent material. A simple example might be one soil formed on schist (a silicate-containing metamorphic rock rich in mica) juxtaposed with a soil formed on serpentine (a ferromagnesian metamorphic rock rich in olivine). More subtle discontinuous lithosequences, such as those on glacial tills, show systematic variation of mineralogical composition or of texture in unconsolidated parent materials.

Topography

Topography, when considered as a soil-forming factor, includes the following: the geologic structural characteristics of elevation above mean sea level, aspect (the compass orientation of a landform), slope configuration (i.e., either convex or concave), and relative position on a slope (that is, from the toe to the summit). Topography influences the way the hydrologic cycle affects earth material, principally with respect to runoff processes and evapotranspiration. Precipitation may run off the land surface, causing soil erosion, or it may percolate into soil profiles and become part of subsurface runoff, which eventually makes its way into the stream system. Erosive runoff is most likely on a convex slope just below the summit, whereas lateral subsurface runoff tends to cause an accumulation of soluble or suspended matter near the toeslope. The conversion of precipitation into evapotranspiration is favoured by lower elevation and an equatorially facing aspect.

Adjacent soils that show differing profile characteristics reflecting the influence of local topography are called toposequences. As a general rule, soil profiles on the convex upper slopes in a toposequence are more shallow and have less distinct subsurface horizons than soils at the summit or on lower, concave-upward slopes. Organic matter content tends to increase from the summit down to the toeslope, as do clay content and the concentrations of soluble compounds.

Often the dominant effect of topography is on subsurface runoff (or drainage). In humid temperate regions, well-drained soil profiles near a summit can have thick E horizons (the leached layers) overlying well-developed clay-rich Bt horizons, while poorly drained profiles near a toeslope can have thick A horizons overlying extensive Bg horizons (lower layers whose pale colour signals stagnation under water-saturated conditions). In humid tropical regions with dry seasons, these profile characteristics give way to less distinct horizons, with accumulation of silica, manganese, and iron near the toeslope, whereas in semiarid regions soils near the toeslope have accumulations of the soluble salts sodium chloride or calcium sulfate.

These general conclusions are tempered by the fact that topography is susceptible to great changes over time. Soil erosion by water or wind removes A horizons and exposes B horizons to weathering. Major portions of entire soil profiles can move downslope suddenly by the combined action of water and gravity. Catastrophic natural events, such as volcanic eruptions, earthquakes, and devastating storms, can have obvious consequences for the instability of geomorphologic patterns.

ORGANISMS

The development of soils can be significantly affected by vegetation, animal inhabitants, and human populations. Any array of contiguous soils influenced by local flora and fauna is termed a biosequence. To return to the climosequence along the Cascade and Sierra Nevada ranges discussed above, the vegetation observed along this narrow foothill region varies from shrubs in the dry south to needle-leaved trees in the humid north, with extensive grasslands in between. In the middle of the precipitation range, transition zones occur in which small groves of needle-leaved trees are interspersed with grassland patches in an apparently random manner. These plant populations represent local flora largely selected by climate. The properties of the soils underlying these plants, however, exhibit differences that do not arise from climate, topography, or parent material but are an effect of the differing plant species. The soils under trees, for instance, are much more acidic and contain much less humus than those under grass, and nitrogen content is considerably greater in the grassland soil. These properties come directly from the type of litter produced by the two different kinds of vegetation.

An opportunity to examine biosequences is often presented by relatively young soils formed from an alluvial parent material. Soils of this kind lying beneath shrubs may be richer in humus and plant nutrients than similar soils found beneath needle-leaved trees. This variation results from differences in the cyclic processes of plant growth, litter production, and litter decay. Organic matter decomposers will feed on stored material in soil if litter production is low, whereas high litter production will permit soil stocks of organic matter to increase, leading to humus-rich A horizons as opposed to the leached E horizons found in soils that form under humid climatic conditions.

Human beings are also part of the biological influx that influences soil formation. Human influence can be as severe as wholesale removal or burial (by urbanization) of an entire soil profile, or it can be as subtle as a gradual modification of organic matter by agriculture or of soil structure by irrigation. The chemical and physical properties of soils critical to the growth of crops often are affected significantly by cultural practices. Among the problems created for agriculture by cultural practices themselves are loss of arable land, erosion, the buildup of salinity, and the depletion of organic matter.

In general, most plants grow by absorbing nutrients from the soil. Their ability to do this depends on the nature of the soil. Depending on its location, a soil contains some combination of sand, silt, clay, and organic matter. The makeup of a soil (soil texture) and its acidity (pH) determine the extent to which nutrients are available to plants.

Soil texture affects how well nutrients and water are retained in the soil. Clays and organic soils hold nutrients and water much better than sandy soils. As water drains from sandy soils, it often carries nutrients along with it. This condition is called leaching. When nutrients leach into the soil, they are not available for plants to use.

An ideal soil contains equivalent portions of sand, silt, clay, and organic matter. Soils across North Carolina vary in their texture and nutrient content, which makes some soils more productive than others. Sometimes, the nutrients that plants need occur naturally in the soil. Othertimes, they must be added to the soil as lime or fertilizer.

Soil pH (a measure of the acidity or alkalinity of the soil)

Soil pH is one of the most important soil properties that affects the availability of nutrients.

- ⦿ Macronutrients tend to be less available in soils with low pH.
- ⦿ Micronutrients tend to be less available in soils with high pH.

Lime can be added to the soil to make it less sour (acid) and also supplies calcium and magnesium for plants to use. Lime also raises the pH to the desired range of 6.0 to 6.5.

In this pH range, nutrients are more readily available to plants, and microbial populations in the soil increase. Microbes convert nitrogen and sulfur to forms that plants can use. Lime also enhances the physical properties of the soil that promote water and air movement.

SOIL ENGINEERING

The use of soil as an engineering material may be said to be as old as mankind itself. Since that time, man has been confronting many types of problems while dealing with soils.

Excellent pavements – Egypt and India much before the Christian Era.

Some earth dams have been used for storage of water in India for more than 2000 years.

During the excavation at the early civilization sites of at Mohenjo-Daro and Harappa in the Indian subcontinent indicate the use of soil as foundation and construction material.

Egyptians used caissons for deep foundations even in 2000 BC.

The hanging gardens at Babylon (Iraq) were also built during the period.

The leaning tower of Pisa was also built around same time. The tower has leaned on one side because of the differential settlement of its base.

In the 17th century, Leonardo da Vinci constructed a number of structures in France, and the London Bridge in England.

The Taj Mahal at Agra is built on masonry cylindrical wells sunk into the soil at close intervals.

The builders were guided by the knowledge and experience passed down from generation to generation.

In 1773, a French engineer Coulomb gave the theory of earth pressure on retaining walls. Coulomb also introduced the concept that the shearing resistance of soil consists of two components – cohesion and friction.

Darcy in 1856 gave the law of permeability. This law is used for the computation of seepage through soils.

In the same year, Stokes gave the law for the velocity of fall of solid particles through fluids. This law is used for determining the particle size.

O-Mohr in 1871 gave the rupture theory for soils. He gave a graphical method of representation of stresses. Popularly known as Mohr's circle, it is extremely useful for determining stresses on inclined planes.

Boussinesq in 1885 gave the theory of stress distribution in a semi-infinite homogeneous, isotropic, elastic medium due to an externally applied load. The theory is used for determining stresses in soils due to loads.

Atterberg in 1911, suggested some simple tests for characterizing the consistency of cohesive soils. These limits are useful for identification and classification of soils.

Prof. Fellenins (Sweden) in 1913 studied the stability of slopes. Swedish circle method for checking the stability of Sweden slopes are popularly used.

The modern era of soil engineering began in 1925, with the publication of the book, Eradbaumechanic, by Karl Terzaghi. He is fittingly called the father of soil mechanics.

His theory of consolidation of soils and the effective stress principle gave a new direction.

Proctor in 1933, did a pioneering work on the compaction of soils.

Taylor worked on the consolidation of soils, shear strength of clays and the stability of slopes.

Casagrande worked on the classification of soils, seepage through earth masses and consolidation.

Skempton did a pioneering work on pore pressure, effective stress, bearing capacity and the stability of slopes.

Mayerhof gave the theories of B.C. of shallow and deep foundations.

Hvorlov did a commendable work on sub-surface exploration and on the shear strength of re-moulded clays.

IMPORTANCE OF SOIL ENGINEERING

Once it is accepted that soil is a structural material, its importance in Civil Engineering becomes paramount. A Geotechnical Engineer should have thorough knowledge of this material of structure as in the case of any other structural material.

Study of Soil Engineering is particularly important in respect of infrastructure development and constructions, viz., highway and airport pavements, foundations and underground structures, retaining walls and embankments and multistorey buildings.

Foundation is considered the most critical part of any structure and it is on its soundness that the stability of the entire structure depends. Since the load bearing capacity of the foundation has a direct relationship with the soil characteristics, the importance of soil investigation should not be underestimated.

FUNDAMENTALS OF SOIL ENGINEERING:

The word 'soil' derives from the Latin word solium, which means, the upper layer of the earth that may be dug or powdered, specifically, the loose surface material of the earth in which plants grow.

The term 'soil' in soil engineering is defined as an unconsolidated material composed of solid particles, produced by the disintegration of rocks. The void space between the particles may contain air, water or both. The solid particles may contain organic matter. The soil particles can be separated by such mechanical means as agitation in water.

A natural aggregate of mineral particles bonded by strong and permanent cohesive forces is called a 'rock'.

Application of laws and principles of mechanics and hydraulics to engineering problems in dealing with soil is usually referred to as Soil Mechanics. The term soil engineering is used to cover a much wider scope implying that it is a practical science rather than a purely fundamental or mathematical one. Hence, Soil Engineering is an applied science dealing with the application of the principles of soil mechanics to practical problems.

It includes site investigation, design and construction of the foundation, earth retaining structures and earth structures.

In the design of any foundation system, the central problem is to prevent the settlements large enough to damage the structure. Just how much settlements to permissible depends on the size, the type and use of the structure, the type of foundation, the source is the subsoil of the settlement, and the location of the structure. In most cases, the critical settlement is not the total settlement but rather the differential settlement, which is the relative movement of the structure.

EMBANKMENT ON SOFT SOIL

Even though a steel storage tank is a flexible structure, a settlement of 1.5 m is too large to be tolerated.

Soil engineering studies show that a very economical solution to the tank foundation problem consists of building on the earth embankment at the site to compress the soft soil, removing the embankment and finally placing the tank on the prepared foundation soil. Such a technique is termed preloading.

FOUNDATION HEAVE

In areas of arid regions, the soils dry and shrink during the arid weather and then expand when moisture becomes available.

Water—rainfall drainage—or from capillary.

When an impervious surface is placed on the surface of the soil, it prevents evaporation. Obviously, the lighter a structure, the more the expanding soil will raise it. Heave problems are commonly associated with light structures such as small buildings, dam spillways and road pavements.

To avoid heave problem first holes are augered into the soil. Steel shells are placed and then concrete base plugs and piles are poured.

Under the building and around the piles an air gap is left, which serves to reduce the amount of heave of the soil (by permitting evaporation) and also to allow room for such heave without disturbing the building.

SOIL AS A CONSTRUCTION MATERIAL:

Soil is essentially the only locally available construction material. Earth has been used for the construction of monuments, tombs, dwellings, transportation facilities and water retention structures.

      a. Earth dam and embankment

      b. Highway pavement

SLOPE AND EXCAVATIONS:

When a soil surface is not horizontal there is a component of gravity tending to move the soil downward.

Stability analysis has to be carried out.

Underground Structures:

Tunnels, shafts and conduits require evaluation of forces exerted by the soil on these structures.

SPECIAL SOIL ENGINEERING PROBLEMS:

Vibrations:

Certain granular soils can be readily densified by vibrations. A building may undergo a considerable settlement due to vibrations – (a) compressors (b) turbines.

Explosions and Earthquakes:

Effects on building of earth waves caused by quarry blasting and other blasting for construction purposes. Similar problems arise as a result of earthquakes.

Frost

Frost heave problems – When in contact with moisture and subjected to freezing temperature, they can imbibe water and undergo a large expansion. Such heave exerts forces large enough to move and crack adjacent structures and can cause serious problems on thawing because of the excess moisture.

The civil engineer designing highways and airfield pavements in frost areas must either select a combination of base soil and drainage that precludes frost heave or design the pavement to withstand the weak soil that occurs in the spring when the frost melts.

Regional Subsidence

Large scale pumping of oil and water from the ground can cause major settlements over a large area.

The first step in minimizing such regional subsidence is to locate the earth material that are compressing as the fluid is removed, and then consider method of replacing the lost fluid.

The interpretation of insufficient and conflicting data, the selection of soil parameters, the modification of a solution, etc., require experience and a high degree of engineering judgement.

While a sound knowledge of soil mechanics is essential for the soil engineer, engineering judgement is usually the characteristic that distinguishes the outstanding soil engineer.

LIMITATIONS OF SOIL ENGINEERING

Over the past century many advances in soil engineering have greatly improved our ability to predict the behaviour of geo-material. However, we still need to maintain skepticism.

Most of our analyses are handicapped by uncertainties introduced by the site exploration and characterization programme. In addition, our mathematical models of soil behaviour are only approximate, and often do not explicitly consider important factors.

The solutions obtained in most cases are for an idealized, hypothetical material, which may not truly represent the actual soil. A good engineering judgement is required for interpreting the results.

2

GEOTECHNICAL PROPERTIES OF SOIL

INTRODUCTION

The civil engineering structures like building, bridge, highway, tunnel, dam, tower, etc. are founded below or on the surface of the earth. For their stability, suitable foundation soil is required. To check the suitability of soil to be used as foundation or as construction materials, its properties are required to be assessed. As per different researchers, assessment of geotechnical properties of subsoil at project site is necessary for generating relevant input data for design and construction of foundations for the proposed structures.

Researchers have stated that proper design and construction of civil engineering structures prevent an adverse environmental impact or structural failure or post construction problems. Information about the surface and sub-surface features is essential for the design of structures and for planning construction techniques. When buildings impose very heavy loads and the zone of influence is very deep, it would be desirable to invest some amount on sub-surface exploration than to overdesign the building and make it costlier. For complex projects involving heavy structures, such as bridges, dams, multi-storey buildings, it is essential to have detail exploration. The purpose of detailed explorations is to determine the engineering properties of the soils for different strata. When the foundations of any structure are constructed on compressible soil, it leads to settlement.

Knowledge of the rate at which the compression of the soil takes place is essential from design consideration. The properties of the soil such as plasticity, compressibility or strength of the soil always affect the design in the construction. Lack of understanding of the properties of the soil can lead to the construction errors. The suitability of soil for a particular use should be determined based on its engineering characteristics and not on visual inspection or apparent similarity to other soils. The loading capability of soil depends upon the type of soil.

Generally, fine grained soils have a relative smaller capacity in bearing of load than the coarser grained soils. Plasticity index and liquid limit are the important factors that help an engineer to understand the consistency or plasticity of clay. Though shearing strength constants at liquid limits but varies for plastic limits for all clays. Permeability influences the civil engineering structures. As per Karsten et al., the shear strength of soils is of special relevance among geotechnical soil properties because it is one of the essential parameters for analyzing and solving stability problems (calculating earth pressure, the bearing capacity of footings and foundations, slope stability or stability of embankments and earth dams).

INTRODUCTION TO SOIL SCIENCE

Soil Science has six well defined and developed disciplines. Scope of soil Science is reflected through these disciplines.

Soil Science

The science dealing with soil as a natural resource on the surface of the earth, including Pedology (soil genesis, classification and mapping) and the physical, chemical and biological and fertility properties of soil and these properties in relation to their management for crop production.

1. **Soil fertility:** Nutrient supplying properties of soil
2. Soil chemistry: Chemical constituents, chemical properties and the chemical reactions
3. **Soil physics:** Involves the study of physical properties
4. **Soil microbiology:** deals with micro organisms, its population, classification, its role in transformations
5. **Soil conservation:** Dealing with protection of soil against physical loss by erosion or against chemical deterioration i.e. excessive loss of nutrients either natural or artificial means.
6. **Pedology:** Dealing with the genesis, survey and classification

Soil can be compared to various systems of human body:

Digestive	-	matters decomposition
Respiratory	-	air circulation & exchange of gases
Circulatory	-	water movement with in the soil system
Excretory	-	leaching out of excess salts
Brain	-	soil clay
Colour	-	soil colour
Height	-	soil depth

Components of Soil (Volume basis)

Mineral matter	-	45%
Organic matter	-	5%
Soil water	-	25%
Soil air	-	25%

Weathering of Minerals

There are many factors which influence the weathering of minerals.

1. Climatic conditions
2. Physical characteristics
3. Chemical and structural characteristics

Climatic Conditions

The climatic condition, more than any other factor tends to control the kind and rate of weathering. Under conditions of low rainfall, there is a dominance of physical weathering which reduces the size and increases the surface area with little change in volume.

The increase in moisture content encourages chemical as well as mechanical changes and new minerals and soluble products are formed.

The weathering rates are generally fastest in humid tropical regions as there is sufficient moisture and warmth to encourage chemical decomposition.

The easily weather able minerals disappear on account of intense chemical weathering and more resistant products (hydrous oxides of Fe and Al) tend to accumulate

Climate controls the dominant type of vegetation which in turn controls the bio chemical reactions in soils and mineral weathering.

Physical Characteristics

i. Differential composition
ii. Particle size and
iii. Hardness and degree of cementation

Chemical and structural characteristics

Chemical: For minerals of given particle size, chemical and crystalline characteristics determine the ease of decomposition. (e.g.) gypsum – sparingly soluble in water, is dissolved and removed in solution form under high rainfall.

Ferro magnesium minerals are more susceptible to chemical weathering than feldspar and quartz

Tightness of packing of ions in crystals: Less tightly packed minerals like olivine and biotitic are easily weathered as compared to tightly packed zircon and muscovite (resistant)

Soil Forming Factors

The soil formation is the process of two consecutive stages.

1. The weathering of rock (R) into Regolith
2. The formation of true soil from Regolith

The evolution of true soil from regolith takes place by the combined action of soil forming factors and processes.

1. The first step is accomplished by weathering (disintegration & decomposition)
2. The second step is associated with the action of Soil Forming Factors

Weathering Factors

Dokuchaiev (1889) established that the soils develop as a result of the action of soil forming factors

$S = f (P, Cl, O)$

Further, Jenny (1941) formulated the following equation

$S = f (Cl, O, R, P, T, ...)$

Where,

Cl	=	environmental climate
o	=	Organisms and vegetation (biosphere)
r	=	Relief or topography
p	=	Parent material
t	=	Time
...	=	Additional unspecified factors

Active Soil Forming Factors

The active soil forming factors are those which supply energy that acts on the mass for the purpose of soil formation. These factors are climate and vegetation (biosphere).

Climate

Climate is the most significant factor controlling the type and rate of soil formation. The dominant climates recognized are:

1. **Arid climate:** The precipitation here is far less than the water-need. Hence the soils remain dry for most of the time in a year.
2. **Humid climate:** The precipitation here is much more than the water need. The excess water results in leaching of salt and bases followed by translocation of clay colloids.
3. **Oceanic climate:** Moderate seasonal variation of rainfall and temperature.
4. **Mediterranean climate:** The moderate precipitation. Winters and summers are dry and hot.

5. **Continental climate:** Warm summers and extremely cool or cold winters.
6. **Temperate climate:** Cold humid conditions with warm summers.
7. **Tropical and subtropical climate:** Warm to hot humid with isothermal conditions in the tropical zone.

Climate affects the soil formation directly and indirectly.

Directly, climate affects the soil formation by supplying water and heat to react with parent material.

Indirectly, it determines the fauna and flora activities which furnish a source of energy in the form of organic matter. This energy acts on the rocks and minerals in the form of acids, and salts are released. The indirect effects of climate on soil formation are most clearly seen in the relationship of soils to vegetation.

Precipitation and temperature are the two major climatic elements which contribute most to soil formation.

Precipitation

Precipitation is the most important among the climatic factors. As it percolates and moves from one part of the parent material to another. It carries with it substances in solution as well as in suspension. The substances so carried are re deposited in another part or completely removed from the material through percolation when the soil moisture at the surface evaporates causing an upward movement of water. The soluble substances move with it and are translocated to the upper layer. Thus rainfall brings about a redistribution of substances both soluble as well as in suspension in soil body.

Temperature

1. Temperature is another climatic agent influencing the process of soil formation.
2. High temperature hinders the process of leaching and causes an upward movement of soluble salts.
3. High temperature favors rapid decomposition of organic matter and increase microbial activities in soil while low temperatures induce leaching by reducing evaporation and there by favour the accumulation of organic matter by slowing down the process of decomposition. Temperature thus controls the rate of chemical and biological reactions taking place in the parent material.

Researcher computed that in the tropical regions the rate of weathering proceeds three times faster than in temperate regions and nine times faster than in arctic.

Organism and Vegetation

Organism

1. The active components of soil ecosystem are plants, animals, microorganisms and man.
2. The role of microorganisms in soil formation is related to the humification and mineralization of vegetation

3. The action of animals especially burrowing animals to dig and mix-up the soil mass and thus disturb the parent material

4. Man influences the soil formation through his manipulation of natural vegetation, agricultural practices etc.

5. Compaction by traffic of man and animals decrease the rate of water infiltration into the soil and thereby increase the rate of runoff and erosion.

Vegetation

1. The roots of the plants penetrate into the parent material and act both mechanically and chemically.

2. They facilitate percolation and drainage and bring about greater dissolution of minerals through the action of CO_2 and acidic substances secreted by them.

3. The decomposition and humification of the materials further adds to the solubilization of minerals

4. Forests - reduces temperature, increases humidity, reduce evaporation and increases precipitation.

5. Grasses reduce runoff and result greater penetration of water in to the parent material.

Passive Soil forming factors

The passive soil forming factors are those which represent the source of soil forming mass and conditions affecting it. These provide a base on which the active soil forming factors work or act for the development of soil.

Parent Material

It is that mass (consolidated material) from which the soil has formed.

Two groups of parent material

i. Sedentary: Formed in original place. It is the residual parent material. The parent material differ as widely as the rocks

ii. Transported: The parent material transported from their place of origin. They are named according to the main force responsible for the transport and re-deposition.

a. By gravity - Colluvial

b. By water - Alluvial, Marine, Locustrine

c. By ice - Glacial

d. By wind – Eolian

Colluvium

It is the poorly sorted materials near the base of strong slopes transported by the action of gravity.

Alluvium

The material transported and deposited by water is, found along major stream courses at the bottom of slopes of mountains and along small streams flowing out of drainage basins.

Locustrine

Consists of materials that have settled out of the quiet water of lakes.

Moraine

Consists of all the materials picked up, mixed, disintegrated, transported and deposited through the action of glacial ice or of water resulting primarily from melting of glaciers.

Loess or Aeolian

These are the wind blown materials.

When the texture is silty - loess; when it is sand - Eolian.

The soils developed on such transported parent materials bear the name of the parent material; viz. Alluvial soils from alluvium, Colluvial soils from Colluvium etc. In the initial stages, however, the soil properties are mainly determined by the kind of parent material.

Endodynamomorphic soils

With advanced development and excessive leaching, the influence of parent material on soil characteristics gradually diminishes. There are soils wherein the composition of parent material subdues the effects of climate and vegetation. These soils are temporary and persist only until the chemical decomposition becomes active under the influence of climate and vegetation.

Ectodynamomorphic Soils

Development of normal profile under the influence of climate and vegetation.

Soil properties as influenced by parent material: Different parent materials affect profile development and produce different soils, especially in the initial stages.

1. Acid igneous rocks (like granite, rhyolite) produce light-textured soils (Alfisols).
2. Basic igneous rocks (basalt), alluvium or Colluvium derived from limestone or basalt, produce fine-textured cracking-clay soils (Vertisols).
3. Basic alluvium or Aeolian materials produce fine to coarse-textured soils (Entisols or Inceptisols).
4. The nature of the elements released during the decaying of rocks has a specific role in soil formation. (e.g.) Si and Al form the skeleton for the production of secondary clay minerals.
5. Iron and manganese are important for imparting red colour to soils and for oxidation and reduction phenomena.
6. Sodium and potassium are important dispersing agents for day and humus colloids.

7. Calcium and magnesium have a flocculating effect and result in favorable and stable soil structure for plant growth.

Relief or Topography

The relief and topography sometimes are used as synonymous terms. They denote the configuration of the land surface. The topography refers to the differences in elevation of the land surface on a broad scale.

Soil formation on flat to almost flat position

On level topographic positions, almost the entire water received through rain percolates through the soil. Under such conditions, the soils formed may be considered as representative of the regional climate. They have normal solum with distinct horizons. But vast and monotonous level land with little gradient often has impaired drainage conditions.

Soil formation on undulating topography

The soils on steep slopes are generally shallow, stony and have weakly- developed profiles with less distinct horizonation. It is due to accelerated erosion, which removes surface material before it has the time to develop. Reduced percolation of water through soil is because of surface runoff, and lack of water for the growth of plants, which are responsible for checking of erosion and promote soil formation.

Soil formation in depression

The depression areas in semi-arid and sub humid regions reflect more moist conditions than actually observed on level topographic positions due to the additional water received as runoff. Such conditions (as in the Tarai region of the Uttar Pradesh) favour more vegetative growth and slower rate of decay of organic remains. This results in the formation of comparatively dark- coloured soils rich in organic matter (Mollisols).

Soil formation and Exposure/ Aspect

Topography affects soil formation by affecting temperature and vegetative growth through slope exposures (aspect}. The southern exposures (facing the sun) are warmer and subject to marked fluctuations in temperature and moisture. The northern exposures, on the other hand are cooler and more humid. The eastern and western exposures occupy intermediate position in this respect.

Time

Soil formation is a very slow process requiring thousands of years to develop a mature pedon. The period taken by a given soil from the stage of weathered rock (i.e. regolith) up to the stage of maturity is considered as time. The matured soils mean the soils with fully developed horizons (A, B, C). It takes hundreds of years to develop an inch of soil. The time that nature devotes to the formation of soils is termed as Pedological Time.

It has been observed that rocks and minerals disintegrate and/or decompose at different rates; the coarse particles of limestone are more resistant to disintegration than those of sandstone. However, in general, limestone decomposes more readily than sandstone (by chemical weathering).

1. The soil properties also change with time, for instance nitrogen and organic matter contents increase with time provided the soil temperature is not high.
2. $CaCO_3$ content may decrease or even lost with time provided the climatic conditions are not arid
3. In humid regions, the H^+ concentration increases with time because of chemical weathering.

Weathering stages in soil formation

Sl. No	Stages	Characteristic
1	Initial	Un weathered parent material
2	Juvenile	Weathering started but much of the originalmaterial still un weathered
3	Virile	Easily weather able minerals fairly decomposed; clay content increased, slowly weather able minerals still appreciable
4	Senile	Decomposition reaches at a final stage; onlymost resistant minerals survive
5	Final	Soil development completed under prevailingenvironments

Soil Forming Processes

The pedogenic processes, although slow in terms of human life, yet work faster than the geological processes in changing lifeless parent material into true soil full of life.

⊙ The pedogenic processes are extremely complex and dynamic involving many chemical and biological reactions, and usually operate simultaneously in a given area.

⊙ One process may counteract another, or two different processes may work simultaneously to achieve the same result.

⊙ Different processes or combination of processes operate under varying natural environment.

The collective interaction of various soil forming factors under different environmental conditions set a course to certain recognized soil forming processes.

The basic process involved in soil formation includes the following.

⊙ Gains or Additions of water, mostly as rainfall, organic and mineral matter to the soil.

⊙ Losses of the above materials from the soil.

⊙ Transformation of mineral and organic substances within the soil.

- ⦿ Translocation or the movement of soil materials from one point to another within the soil. It is usually divided into
  1. Movement of solution (leaching) and
  2. Movement in suspension (eluviation) of clay, organic matter and hydrous oxides

GEOTECHNICAL PROPERTIES OF SOILS

Different geotechnical property of soils has different influence on the civil engineering structures. They also depends upon each other. The properties are discussed as under:

Specific Gravity

Specific gravity is the ratio of the mass of soil solids to the mass of an equal volume of water. It is an important index property of soils that is closely linked with mineralogy or chemical composition and also reflects the history of weathering. It is relatively important as far as the qualitative behavior of the soil is concerned and useful in soil mineral classification, for example iron minerals have a larger value of specific gravity than silicas. It gives an idea about suitability of the soil as a construction material; higher value of specific gravity gives more strength for roads and foundations. It is also used in calculation of void ratio, porosity, degree of saturation and other soil parameters. Typical values of specific gravity are given in Table 1.

Table 1: Typical values of specific gravity of soil

Type of soil	Specific gravity
Sand	2.65-2.67
Silty sand	2.67-2.70
Inorganic clay	2.70-2.80
Soil with mica or iron	2.75-3.00
Organic soil	1.00-2.60

Density Index

The degree of compaction of fine grained soils is measured in relation to maximum dry density for a certain compactive effort, like 90% of light compaction density or proctor density. But in case of coarse grained soils, a different sort of index is used for compaction. Depending upon the shape, size, and gradation of soil grains, coarse grained soils can remain in two extreme states of compaction, namely in the loosest and densest states. Any intermediate state of compaction can be compared to these two extreme states using an index called relative density or density index. The soil characteristics based on relative density are shown in Table 2.

Density index is expressed in percent and is defined as the ratio of the difference between the void ratio of a cohesionless soil in the loosest state and any given void ratio to the difference between its void ratios in the loosest and the densest states. It is a measure of the degree of compactness, and the stability of a stratum.

Table 2: The soil characteristics based on relative density

Relative density (%)	Soil compactness	Angle of shearing Resistance (%)
0-15	Very loose	<28
15-35	Loose	28-36
35-65	Medium	30-60
65-85	Dense	36-41
85-100	very dense	>41

Consistency Limits

The consistency of a fine-grained soil is largely influenced by the water content of the soil. A gradual decrease in water content of a fine-grained soil slurry causes the soil to pass from the liquid state to a plastic state, from the plastic state to a semi-solid state, and finally to the solid state. The water contents at these changes of state are different for different soils. The water contents that correspond to these changes of state are called the Atterberg limits. The water contents corresponding to transition from one state to the next are known as the liquid limit, the plastic limit and the shrinkage limit.

The liquid limit of a soil is the water content, expressed as percentage of the weight of the oven dried soil, at the boundary between the liquid and plastic states of consistency of the soil. The soil has negligibly small shear strength. The plastic limit of a soil is the water content, expressed as a percentage of the weight of oven dried soil, at the boundary between the plastic and semi-solid states of consistency of the soil.

The plastic limit for different soils has a narrow range of numerical values. Sand has no plastic stage, but very fine sand exhibits slight plasticity. The plastic limit is an important soil property. Earth roads are easily usable at this water content. Excavation work and agricultural cultivation can be carried out with the least effort with soils at the plastic limit. Soil is said to be in the plastic range when it possesses water content in between liquid limit and plastic limit. The range of the plastic state is given by the difference between liquid limit and plastic limit and is defined as the plasticity index. The plasticity index is used in soil classification and in various correlations with other soil properties as a basic soil characteristic.

Particle Size Analysis

The percentage of different sizes of soil particles coarser than 75 μ is determined by sieve analysis whereas less than 75 μ are determined by hydrometer analysis. Based on the particle size analysis, particle size distribution curves are plotted. The particle size distribution curve (gradation curve) represents the distribution of particles of different sizes in the soil mass. It gives an idea regarding the gradation of the soil i.e. it is possible to identify whether a soil is well graded or poorly graded. In mechanical soil stabilization, the main principle is to mix a few selected soils in such a proportion that a desired grain size distribution is obtained for the design mix. Hence for proportioning the selected soils, the grain size distribution of each soil is required to be known.

Researchers explained that the grain size analysis is widely used in classification of soils. The data obtained from grain size distribution curves is used in the design of filters for earth dams and to determine suitability of soil for road construction, air fields, etc. Researchers stated that the particle size of sands and silts has some practical value in design of filters and in the assessment of permeability, capillarity, and frost susceptibility. Very relevant and useful information may be obtained from grain size curve such as (i) the total percentage of larger or finer particles than a given size and (ii) the uniformity or the range in grain-size distribution.

Researchers found that particle-size is one of the suitability criteria of soils for roads, airfield, levee, dam, and other embankment construction. Information obtained from particle-size analysis can be used to predict soil-water movement, although permeability tests are more generally used. The susceptibility to frost action in soil, an extremely important consideration in colder climates, can be predicted from the particle-size analysis.

Very fine soil particles are easily carried in suspension by percolating soil water, and under drainage systems are rapidly filled with sediments unless they are properly surrounded by a filter made of appropriately graded granular materials. The proper gradation of this filter material can be predicted from the particle-size analysis. Particle-size of the filter materials must be larger than the soil being protected so that the filter pores could permit passage of water but collect the smaller soil particles from suspension.

As per researchers, the sand shape whether rounded, subrounded, or angular will affect the shearing strength of soil. Angular grains provide more interlock and increased shear resistance. The gradation and size of the sand affect the shear resistance. Well-graded materials provide more grain to grain area contact than poorly graded materials. Porosity and spaces available for clay within the sand is an important while considering the mixtures of clays and sands.

Compaction

Soil compaction is one of the ground improvement techniques. It is a process in which by expending compactive energy on soil, the soil grains are more closely rearranged. Compaction increases the shear strength of soil and reduces its compressibility and permeability.

Researchers explained that when an earth dam is properly compacted, the shear strength of the material is increased and dam becomes more stable. Since the soil becomes dense, its permeability gets decreased. The decrease in the permeability of the dam decreases the seepage loss of the water stored. The settlement of the dam also decreases due to the increase in the density of the materials.

According to some researchers, compaction of soils increases the density, shear strength, bearing capacity but reduces their void ratio, porosity, permeability and settlements. The results are useful in the stability of field problems like earthen dams, embankments, roads and airfields. The moisture content at which the soils are compacted in the field is controlled by the value of optimum moisture content determined by the laboratory proctor compaction test. The compaction energy applied in the field is also controlled by the maximum dry density determined in the laboratory.

Consolidation

When a soil layer is subjected to compressive stress due to construction activities, it undergoes compression. The compression is caused by rearrangement of particles, seepage of water, crushing of particles, and elastic distortions. Settlement of a structure is analyzed for three reasons: appearance of structure, utility of the structure, and damage to the structure. The aesthetic view of a structure can be spoiled due to the presence of cracks or tilt of the structure caused by settlement.

Settlement caused to a structure can damage some of the utilities like cranes, drains, pumps, electrical lines etc. Further settlement can cause a structure to fail structurally and collapse. Settlement is the combination of time-independent (e.g. immediate compression) and time-dependent compression (called consolidation).

The main aim of a consolidation test is to obtain soil data which are used in predicting the rate and amount of settlement of structure founded on clay primarily due to volume change of the clay. The information obtained for foundations resting on clay are:

    i. Total settlement of foundation under any given load,

    ii. Time required for total settlement due to primary consolidation,

    iii. Settlement for any given time and load,

    iv. Time required for any percentage of total settlement or consolidation, and

    v. Pressure due to which soil already has been consolidated/compressed.

Researchers explained that lowering of water-table or dewatering is probably the best known cause of massive settlement. When submerged, soil particles are subjected to buoyancy. Upon dewatering, the buoyancy is removed and the apparent increase in pressure results in consolidation, even though there is no increase in external load.

Vibrations can also have a densification effect on soils and lead to subsequent settlement. The effects can be severe when the vibration frequency matches the soil's natural frequency. Soils often fail and settle disastrously as a result of earthquakes.

Devastating landslides are often one of the results of such occurrences. Of the three phases of soil, only the solid phase controls the resistance to compression and shear. Water, present in a moist soil is highly incompressible but as a liquid, is not capable of resisting shear loads. Air, present in unsaturated soils, will not support compression or shear loads.

Researchers stated that in a saturated soil, compression will be primarily caused by expulsion of water out of the soil voids. Under the influence of an externally applied load, the expulsion of water from the voids is highly dependent on the permeability of the medium. The extremely low permeability in the case of clay leads to a slow void contraction.

The compression of saturated, low permeability layers under a static pressure is known as consolidation. The consolidation rate depends on the compressibility of the soil (rate of decrease in volume with stress) and soil permeability, which in turn, is dependent on the viscosity of the liquid. An increase in temperature increases the consolidation rate but does not affect total amount of consolidation.

Based on the study, researchers found that among other reasons, the effect of the railway as one of the big contributors to the settlement of the Little Hagia Sophia Mosque (Church *of* St. Sergius *and* Bacchus) – Istanbul, Turkey. They found that the railway, which was operational for 50 years at 5 m away from the mosque, caused bricks to fall from the nearby wall when trains were passing by.

The influence of the railway can increase the settlement with weak soil and high water level. Researchers also found that settlement is triggered by earthquakes, frequent changes in underground water level caused by the river base change, changes of ground water level due to surrounding drainage system and constant vibrations generated by the adjacent motorway and railway.

Researchers carried out settlement study for a Institutional Building located in South Goa, India, which developed cracks when the construction had reached till the plinth beam level. It was found that some foundations were located above the natural ground at a depth of 2 m in unconsolidated filled up ground of an abandoned laterite stone quarry, where SPT (Standard penetration Test) was found to be less than 12, which resulted for differential settlement.

This differential settlement was observed towards the front left corner of the Building which was lying on the filled up ground. The differential settlement led to cracks in the plinth beam and Foundation Concrete.

Permeability

The amount, distribution, and movement of water in soil have an important role on the properties and behavior of soil. The engineer should know the principles of fluid flow, as groundwater conditions are frequently encountered on construction projects. Water pressure is always measured relative to atmospheric pressure, and water-table is the level at which the pressure is atmospheric. Soil mass is divided into two zones with respect to the water-table:

   i. Below the water-table (a saturated zone with 100% degree of saturation) and
   ii. Just above the water-table (called the capillary zone with degree of saturation < 100%).

Data from field permeability tests are needed in the design of various civil engineering works, such as cut-off wall design of earth dams, to ascertain the pumping capacity for dewatering excavations and to obtain aquifer constants. The permeability of soils has a decisive effect on the stability of foundations, seepage loss through embankments of reservoirs, drainage of subgrades, excavation of open cuts in water bearing sand, and rate of flow of water into wells.

Researchers explained that water flowing through soil exerts considerable seepage forces, which have direct effect on the safety of hydraulic structures. The rate of settlement of compressible clay layer under load depends on its permeability. The quantity of stored water escaping through and beneath an earthen dam depends on the permeability of the embankment and the foundation respectively.

The rate of drainage of water through wells and excavated foundation pits depends on the coefficient of permeability of the soils. Shear strength of soils also depends indirectly on its permeability, because dissipation of pore pressure is controlled by its permeability. According to U. S. Bureau of Reclamation, soils are classified as (i) Impervious: k (coefficient of permeability) less than 10^{-6} cm/sec, (ii) Semi pervious: k between $^{6\text{-}}10$ to $^{4\text{-}}10$ cm/sec (iii) Pervious: k greater than $^{4\text{-}}10$ cm/sec.

The Hsinchu is located from north to south along the west coastal plain of Taiwan. Taiwan is a seismically active region and has governing seismic design criteria similar to those used in the International Building Code (IBC). At the foundation construction site, different layers were found at different depth like fill (soft, silty clay with variable amounts of sand, gravel, and organic material) clay (medium stiff to stiff, silty clay) Gravel/Cobble.

The hydraulic conductivity (permeability) varied accordingly. The use of permanent drainage systems under the floor slab to draw down the groundwater table allowed the buildings to be supported on the more cost effective shallow footings and slab-on-grade floors.

SHEAR STRENGTH

The shear resistance of soil is the result of friction and the interlocking of particles and possibly cementation or bonding at the particle contacts. The shear strength parameters of soils are defined as cohesion and the friction angle. The shear strength of soil depends on the effective stress, drainage conditions, density of the particles, rate of strain, and direction of the strain. Thus, the shearing strength is affected by the consistency of the materials, mineralogy, grain size distribution, shape of the particles, initial void ratio and features such as layers, joints, fissures and cementation. The shear strength parameters of a granular soil are directly correlated to the maximum particle size, the coefficient of uniformity, the density, the applied normal stress, and the gravel and fines content of the sample. It can be said that the shear strength parameters are a result of the frictional forces of the particles, as they slide and interlock during shearing. Soil containing particles with high angularity tend to resist displacement and hence possess higher shearing strength compared to those with less angular particles.

Different researchers explained that the capability of a soil to support a loading from a structure, or to support its overburden, or to sustain a slope in equilibrium is governed by its shear strength. The shear strength of a soil is of prime importance for foundation design, earth and rock fill dam design, highway and airfield design, stability of slopes and cuts, and lateral earth pressure problems. It is highly complex because of various factors involved in it such as the heterogeneous nature of the soil, the water-table location, the drainage facility, the type and nature of construction, the stress history, time, chemical action, or environmental conditions.

As per some researchers, confining pressures play the significant role in changing the behavior of soils in deep foundations. Similarly in high rise earth dams, the confining pressures are of very high magnitude. Triaxial test is the only test to simulate these confining pressures. For short-term stability of foundations, dams and slopes, shear strength parameters for unconsolidated undrained or consolidated undrained conditions are used, while for long term stability shear parameters corresponding to consolidated drained conditions give more reliable results.

Researchers found that the friction angle is high for a sandy soil than its cohesion and *vice versa* for clayey soil. Researchers also in their study concluded that there is a general increase in cohesion with clay content. As more clay is introduced into the sandy materials, the clay particles fill the void spaces in between the sand particles and begin to induce the sand with interlocking behavior.

Hence, clayey sand soils are expected to exhibit low cohesion whereas the cohesion increases with high clay content.

Researchers observed that the mineralogy can have a major role in the shearing strength capacity of clays. The cementation between particles can either be due to a chemical bond or physicochemical bond. Swelling and shrinkage in expansive soils are of two extreme opposite effects on the shearing strength. The shear strength is generally low for fully expanded clay while dry shrinking clay is capable of developing higher cohesion and angle of internal friction. The study indicated that choosing the appropriate mix or using appropriate quantity of clay, can help to achieve required shear strength. Very moist clay-sand mixture showed steep drop in both cohesion and angle of internal friction when the clay content is high.

Cohesion is mainly due to the intermolecular bond between the adsorbed water surrounding each grain, especially in fine-grained soils. As per researchers, the soils with high plasticity like clayey soils have higher cohesion and lower angle of shearing resistance. Conversely, as the soil grain size increases like sands, the soil cohesion decreases.

SOIL PROPERTIES AND SAMPLE PREPARATION

In this research, one kind of natural soil, i.e., limestone crushed stone as natural aggregate (NA), and three kinds of anthropogenic soils, i.e., recycled concrete aggregate (RCA), fly ash and bottom ash mix (BS) and blast furnace slag (BFS), were used in laboratory tests (Fig. 1). A brief description of each material is presented below. Additionally, a unified granulometric composition was created for each soil. Moreover, the soil fractions for a given soil type were mixed to get the grain size distribution curve common for each soil type.

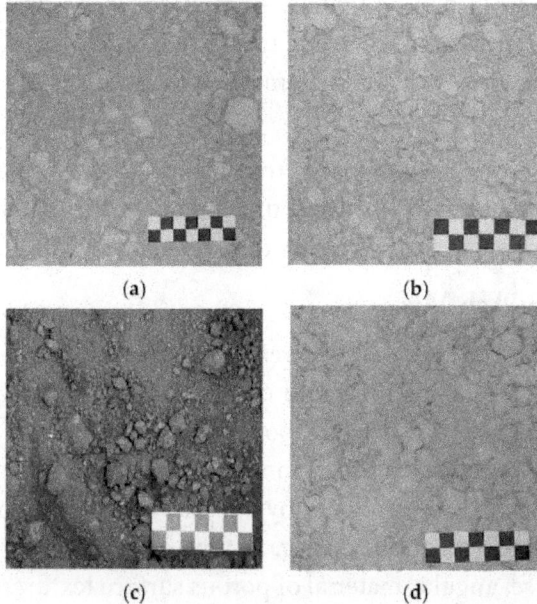

(a)   (b)

(c)   (d)

Fig. 1. Materials tested: (a) crushed limestone, (b) recycled concrete aggregate, (c) fly ash and bottom ash mix, (d) blast furnace slag (the side of the scale in the form of a chessboard has a length of 1 cm).

Crushed Limestone

Crushed limestone is a form of construction NA, typically produced by mining rock deposit and breaking the removed rock down to the desired size using a crusher. NA is one of the most accessible natural resources. Its advantage is that it can be crushed and sized to meet most specifications. This material is clean, angular, and binds well with cementing mixtures. A uniform lithological composition can be maintained with little or no selective quarrying in many areas. The specific gravity of NA is equal to 2.67.

Recycled Concrete Aggregate

As a result of the separation process and later crushing of concrete products, recycled concrete aggregates (RCA) are produced. Despite differences during the tests, factors such as bearing capacity or shear strength have shown an excellent performance of RCA, meaning that it can be applied as a pavement subbase. As an artificial aggregate, RCA characterizes another structure composition in comparison with natural aggregates.

RCA contains portions of the cement matrix that consists of anhydrous cement and hydration products, which form a porous microstructure. The high water absorption of the RCA, especially outer layers called 'attached mortar,' may be the reason for lower mechanical properties when compared to natural aggregates.

RCA in this research was taken from a building demolition site in Warsaw, by the skid-mounted impact crusher. The strength class properties of the construction concrete, made from Portland cement, were estimated at the level from C16/20 to C30/35, based on the data obtained from building plans. Then, the material was fractionated using a mechanical shaker and divided into several types, each one composed of sieved fractions. In this study, only one type with grain diameter dimensions 0–20 mm was investigated in the laboratory.

The aggregates were 99% composed from broken cement concrete, the rest being glass and brick (Σ(Rb, Rg, X) < 1% m/m), under standard, and contained no asphalt or tar elements. The specific gravity of RCA in this study is equal to 2.54.

Fly Ash and Bottom Ash Mix

Fly ash and bottom ash (BS) are the solid residue byproducts produced by coal-burning electric utilities. They are mainly disposed of together as waste in utility disposal sites, where a typical disposal rate for fly ash is about 80% and for bottom ash is about 20%. Fly ash is the finely divided residue resulting from the combustion of powdered coal, which is transported from the firebox to the boiler by flue gas, whereas bottom ash is a byproduct of burning coal at the thermal power plant. Bottom ash particles are much coarser than the fly ash. It is a coarse, angular material of porous surface texture, predominantly sand-sized. The chemical composition of both materials is similar, but bottom ash typically contains a greater quantity of carbon. It exhibits as well high shear strength and low compressibility.

In Poland, approximately 50 million tons of hard coal and 60 million tons of brown coal are used annually to generate electricity. When burning such an amount of coal, vast amounts of slag, ashes, and fly ash are generated. Each year, about 24 million tons of this waste are produced. The amount of fly ash produced over the year, for example, has changed over the last 9 years, ranging from a maximum of 4.6 million tons in 2012 down to 3.3 million tons in 2016 and 2017.

The combustion waste from energy generation is primarily used in the production of construction materials (including cement), the construction of roads, and mining. BS is a material that finds its application in road construction. The demand for this material increases year by year due to the increase in communication investments, plans to expand the motorway network in Poland, and a change in the design of communication routes in cities that are currently carried out on earth embankments instead of using reinforced concrete flyovers.

The usefulness of BS depends on the following features: structure, degree of sintering, and content of ingredients. Its basic characterization for application in the improved substrate, i.e., reinforcing, frost-resistant, and drainage (filtration) layers, are high internal friction angle, low bulk density, as well as high filtration at every operating time.

BS used in this research was supplied by the Siekierki Cogeneration Plant from Warsaw, which is a part of the power station PGNiG TERMIKA. Byproducts of combustion (ash and slag) at the PGNiG TERMIKA plants are regularly transferred to recipients possessing the appropriate licenses. They are used in the following areas: road construction, in the production of construction materials, and as leveling material. The ash and slag generated in large volumes during the peak heating period are stored and then recovered in the summer.

The fly ash and bottom ash were first dried for 24 h and brought to room temperature. Next, they were mixed together in the required proportions, and their geotechnical characteristics were investigated. The specific gravity of AF + BA in this study is equal to 2.03.

Blast Furnace Slag

Blast furnace slag (BFS) is a byproduct of the steel manufacturing industry with a long tradition of use in different areas of civil engineering. BFS is formed when iron ore or iron pellets, coke, and a flux (either limestone or dolomite) are melted together in a blast furnace. When the metallurgical smelting process is complete, the lime in the flux has been chemically combined with the aluminates and silicates of the ore and coke ash to form a nonmetallic product called blast furnace slag.

During the period of cooling and hardening from its molten state, BFS can be cooled in several ways to create any of several types of BF slag products. In most cases, the slag is cooled by water, dried, and ground to a fine powder. This ground granulated blast furnace slag (GGBFS) contains over 95% glass and has high reactivity. When cooling takes place more slowly, the glass fraction decreases, leading to a significant reduction of the reactivity when crystalline minerals are present.

Granulated blast furnace slag has been applied in the cement industry for over 100 years. Nowadays, slag is widely used in the cement and concrete industry, and over 16 million tons per year, granulated slag is produced by the European steel industry. In this study, examined BFS has a hydraulic property, and there is no risk of alkali-aggregate reaction. Because of the potent latent hydraulic property that results from fine grinding, BFS is applied in products such as Portland cement. When blended with cement, GGBFS becomes Portland blast furnace slag cement (PSC) with the same properties as ordinary (Portland) cement. PSC is a mixture of ordinary Portland cement and not more than 65 wt % of granulated slag.

It is generally recognized that the rate of hardening of slag cement is slower than that of ordinary Portland cement during the first 28 days, but after that, increases so that, at 12 months, the strengths become close to or even exceed those of Portland cement. The advantages of this blast furnace slag cement, such as briefly mentioned increasing strength over long periods, low heating speed when reacting with water, and high chemical durability, are put to practical use in a broad range of fields, including the construction of ports and harbors and other significant civil engineering works.

BFS in this research was obtained from the Lafarge Cement SA cement plant, created in the blast furnace process. One type gained as a result of sieve analysis, with a grain diameter of 0–20 mm, was used in the tests. The specific gravity of BFS in this study is equal to 2.58.

STATIC CBR TESTS

We performed CBR tests to determine the engineering properties of the anthropogenic soils used in road applications. The CBR test is a penetration test applied to evaluate the subgrade strength of roads and pavements. In this test, a standard piston, with a diameter of 50 mm, penetrated the soil at the standard rate of 1.25 mm/minute. The pressure up to penetration of 2.5 mm was measured, and its ratio to the bearing value of a standard crushed rock was termed as CBR.

Laboratory CBR tests were conducted to the Proctor method, according to the American Society for Testing and Materials (ASTM) standard. The selected mode is characterized by the use of a 2.5 kg hammer and a larger mold of 150 mm diameter and 120 mm height, with a volume of 2.2 dm^3. A 3-layer Proctor test was performed, with 56 blows to each layer. This procedure creates constant energy of compaction, whose level is equal to 0.59 J/cm^3. We also did the vibratory compaction tests with the use of a vibratory compaction hammer. The compaction was conducted in three layers, and 8 s excitation on each layer was applied. The energy of compaction corresponds with the Proctors test energy of compaction. The preliminary tests highlighted though the maximum dry density and optimum moisture content (OMC).

CYCLIC CBR TESTS

Another mechanical property analysis was the cyclic CBR test. The cCBR tests were conducted on samples prepared in CBR mold by applying the cCBR test procedure. The idea behind the cCBR test comes from the CBR test popularity and uncomplicated test procedure.

The principles of this test during the first step were the same as for the usual CBR test. Each specimen was loaded by force of 0.05 kPa in order to keep in contact with the plunger from the beginning of the test. The test started under standard conditions, i.e., penetration velocity of 1.27 mm/min (the frequency equals to 0.0084 Hz), the penetration depth of 2.54 mm in the first cycle. When the desired plunger penetration was reached, the unloading phase was performed to stress equal to 10% of the maximal stress noted on 2.54 mm penetration. The first loading cycles consisted of the loading and unloading phases. The next cycles were performed to the maximal and minimal stress obtained in the first cycle. The number of cycles was determined by the percentage of the plastic displacements in one cycle (usually 50 to 100 cycles). The test can be considered complete when, in one cycle, less than 1% of noted displacements are plastic.

Hence, in summary, the cCBR test is aimed to simulate displacements of subgrade surface by constant force application and to observe plastic displacements behavior under cyclic loading.

MODIFIED OEDOMETER TESTS

The oedometer consolidation test was adopted in this study for the determination of the compressibility of the tested materials when subjected to vertical loads. The results were used to calculate and estimate the compression index C_c and preconsolidation pressure p'_c. The compression index is the parameter that can be applied to settlement calculations.

In this test, compacted soil specimens were loaded axially in constant stress steps until the primary consolidation ceased. The important aim of modified oedometer tests was to obtain compressibility characteristics of the tested materials in the moisture content in which the samples were compacted. The oedometric tests of our specimens took place in the Proctor cylinder (d = 150 mm, h = 120 mm), which supports no lateral movements of the soil. The following sequence of loading step were applied: 12.5, 25, 50, 100, 200, 400, 800, 1600 kPa. Each increment of loading was held constant for 100 s. This time procedure was chosen to get the compression characteristics and to avoid any additional soil deformation due to time-effect concerning the grain crushing, which is herein called secondary compression. Therefore, in this study, the compression curves only characterized the skeleton compression in first phase of the loading.

SOIL GRADATION CURVE

The four types of soil tested in this article were sieved, and then the standard soil gradation curve 0–20 mm was composed based on the weight share. The gradation curve fulfills the requirements of Polish, English, and American codes, and the soil with such gradation can be used as a road subbase material. The common gradation curve is presented in Fig. 2.

The coefficient of curvature (C_c) and coefficient of uniformity (C_U) were calculated in order to classify the shape of the grading curve of tested types. The C_U value is equal to 30.0, and C_C is equal to 0.53.

The soil was, therefore, classified as well-graded, according to Eurocode 7. The fractions composition of the tested gradation curve has led to recognizing the soil as gravelly sand (grSa), and as well-graded sand or gravelly sand (SW), according to Unified Soil Classification System (USCS).

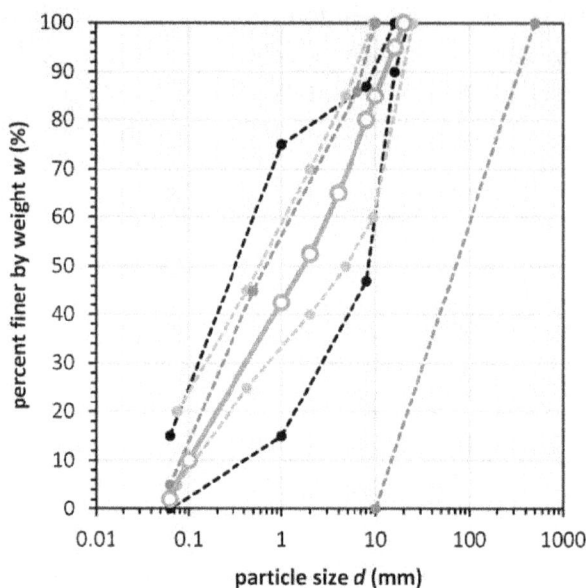

Fig. 2. Soil gradation curve for Natural Aggregate (NA) and Anthropogenic Soils (AS) tested in this study (solid green line, the tested soil material was first sieved into fractions, and then the universal soil grain composition was created for each kind of soil), the red dashed line-British standards requirement, blue dashed line-American standards requirement, black sashed line Polish standards requirement.

OPTIMUM MOISTURE CONTENT

The compaction of four soil types consists of the Proctor test and vibratory hammer tests. The test results show high dependence of compaction test type on the test results. The soil tested with the Proctor method had an inconsistent characteristic, and no apparent impact of moisture content can be seen. The NA reaches the highest density in moisture content equal to 6.52%, which corresponds to the saturation ratio equal to 0.94, and the maximal dry density is equal to 2120.0 kg/m^3.

The gravelly sands or sandy gravels usually have a higher dry density in low moisture contents, and with an increase of water content, the density drops. The compaction curve convex is upwards. When the effect of matric suction decreases due to increased moisture, the dry density rises, which we can observe in the case of NA.

The anthropogenic soils behave quite differently. For example, the RCA has the highest density in the lowest tested moisture content. This conclusion can be drawn based on both vibro-compaction and Proctor compaction tests. The possible reason for that is the grain roughens, which is hard to estimate. Based on the tests performed. concerning the interface effects between RCA particles, it was concluded that between RCA grains exist very low values of tangential stiffness. The authors stated that RCA grains have low hardness and soft nature, which contributed to the test results.

The surface smoothening and debris production impact the surfaces and result in complex effects of preshearing. The grain shearing shows a reduction of surface roughness and damage to micro-asperities. This behavior indicates that the RCA grain surface can undergo changes that can lead to different compaction characteristics. The moisture content and matric suction impact result in a decrease of dry density, and even in the state of high moisture content, the RCA does not have a higher dry density than in air-dried conditions.

The BS compaction curve shows that this material in opposite to two previous types can reach a high moisture content at which the maximum dry density is observed. The fly ash and bottom ash mix in such conditions have the lowest dry density from all four soil types tested in this article. The reason for that is low specific gravity, which is caused by high internal porosity grains in the "popcorn-like" shapes. What is more, the dry density in air-dried conditions has the lowest value, which differs this type of soil from the rest of the tested ones. The intermediate moisture content characterizes with moderate dry density. BS reaches dry density equal to 1300.9 kg/m³ in optimum moisture content equal to %22.4.

Similar to previous cases, the vibro-compaction gives higher dry density results in comparison to the Proctor method. The moisture content in such conditions indicates that the fly ash and bottom ash mix is in near full saturation state (S$_r$ Hd 0.94). Therefore, during the field compaction, BS should be in "flushed" conditions to obtain the highest degree of densification. The BFS compaction has the compaction characteristics close to RCA, but in air-dried conditions, the dry density is significantly lower in comparison to the dry density in OMC.

The compaction curves show proper consistency with the vibro test method. The noncohesive soils have lower susceptibility to the water content than the cohesive soils. This property can be observed as an absence of a typical compaction curve characteristic. The Proctor method was proven to be a less-accurate method for OMC estimation due to the high inconsistency of the test results. The inconsistency is the result of the soil properties. Consequently, we were resigned to use compaction curves to estimate the soil OMC.

STATIC CBR TEST RESULTS

The CBR values were calculated for 2.54 plunger penetration. The highest CBR values for tested soils were observed for vibratory compaction in comparison with the Proctor compaction. The CBR for NA reached 162.4% in optimum moisture content. The RCA has CBR equal to 147.7% in optimum moisture content as well. Lower CBR values were observed in the case of BS (maximum CBR = 42.1%) and BFS (maximum CBR = 93.6%); nevertheless, the top CBR values were observed in OMC. It indicates that the anthropogenic soils are obeying the same rule as NA, where the highest bearing capacity is observed in the OMC conditions. The relation between the anthropogenic soil parameters can be evaluated using the Pearson correlation rang, which gives information about the linear relationship between the soil properties and the CBR value. The results of the calculations are presented in Table 3.

Table 3. Pearson correlation analysis result for CBR value

	Dry Density $ñ_d$ (kg/m³)	Specific Gravity S_G (–)	Compaction Energy E_C (J/cm³)	California Bearing Ratio (CBR) (%)
Moisture m (%)	-0.565	-0.617	0.0997	-0.357
	0.00033	0.00006	0.563	0.0323
Dry Density $ñ_d$ (kg/m³)		0.987	0.162	0.541
		0.0000	0.344	0.0006
Specific Gravity S_G (–)			0.162	0.518
			0.346	0.0012
Compaction Energy E_C (J/cm³)				0.373
				0.02504

The results of the calculations indicate that there exists a significant relationship between CBR parameter change and the physical properties (*p*-value less than 0.050). However, the correlation is rather weak since the correlation coefficient is less than 0.541, in case of dry density (the correlation coefficient equal to 1 or -1 means perfect linear between two variables).

The underlined number stated for linear correlation as the coefficient of determination R^2, the not underlined numbers are the *p*–values where for the *P* greater than 0.050, there is no significant relationship between two variables.

ROLE OF CLAY IN GEOTECHNICAL ENGINEERING

Studies on soil behavior that do not consider the physico-chemical and microstructural properties of clay soils may be missing important information regarding the soil's physical and mechanical properties. This is because most physical and mechanical behaviors can be explained by the soil's physico-chemical and microstructural properties. In general, clay is an unwanted material because it creates significant engineering problems. Unlike other minerals of the same size, clay forms mud when mixed with water. Clay has plasticity and can be shaped into dough, and when cooked it turns into a solid with great strength increments. Clay generally shows a volume increase when wet, and when it is dried, its volume decreases, which creates many cracks.

Physical and mechanical behavior of clay

In geotechnical engineering, it is important to identify a clay type, as the type directly affects the important properties of clay, such as Atterberg's limits, hydraulic conductivity, swelling-shrinkage, settlement (compression) and shear resistance. Atterberg's limits, known as consistency limits, define the relationship between ground particles and water and the state of the soil relative to varying water contents. The clay-water mixture shows a total volume reduction, which is equivalent to the volume of water lost around the liquid and plastic limits, as the clay transitions from liquid to dry, and if the decrease in water content continues, no reduction in volume is observed. This limit value is called the shrinkage limit.

Therefore, the shrinkage limit is the moisture content at which the soil volume will not reduce further if the moisture content is reduced. The plastic limit is the moisture content at which the soil changes from a semisolid to a plastic (flexible) state. The liquid limit is the moisture content at which the soil changes from a plastic to a viscous fluid state. In geotechnical engineering, the liquid and plastic limits are commonly used.

Hydraulic conductivity properties of clay

Water is a problem in geotechnical engineering, such as water in voids in the ground mass, flowing in pores, or in the pressure or stress that water creates in the pores. Clay plays an important role in the emergence of water problems, especially on fine soils, and these problems include permeability, shear resistance, setting and swelling problems. In addition, capillarity, freezing and infiltration can be additional issues. Structures built on clay and slope stability are particularly problematic when affected by water. Dams and dikes also cause the destruction of structures without leakage and piping.

Therefore, it is necessary to estimate the quantity of underground seepage under various hydraulic conditions to investigate problems that involve pumping water for underground construction and for stability analyses of earthen dams and earth-retaining structures that are subject to seepage forces. The hydraulic conductivity coefficient commonly used in geotechnical engineering is also used for permeability. Hydraulic conductivity is a property that expresses how water flows in the soil.

Soils are permeable due to the existence of interconnected voids, through which water can flow from the points of high energy to the points of low energy. Fluid viscosity, pore-size distribution, grain-size distribution, void ratio, roughness of particles and the degree of soil saturation affect the hydraulic conductivity of soils. Clay soil has electrical ions, so the hydraulic conductivity of clays affects the ionic concentration and thickness of water layers held to the clay particles. Table 4 provides the typical values for soils. The hydraulic conductivity value of soils determines the constant head test (for coarse soils) and the falling head test (for fine-grained soils).

Table 4. Hydraulic conductivity of soils

Soil type	k (cm/s)
Clean gravel	100–1.0
Coarse sand	1.0–0.01
Fine sand	0.01–0.001

Soil type	k (cm/s)
Silty clay	0.001–0.00001
Clay	<0.000001

Swelling-shrinkage behavior of clay

The effect of swelling-shrinkage on fine-grained soils is often seen as a problem in geotechnical engineering applications. Shrinkage behavior in clay soils is effective in reducing the strength in a slope and a foundation's bearing capacity. Shrinkage is usually visible from evaporation in dry climates, reduction of groundwater and sudden arid periods. Swelling can be seen due to rising water. These volume changes are harmful to heavy construction and road coverings. Swelling occurs when the inflation pressure is greater than the pressure from the covering or structure. The material damage from the swelling-shrinkage of soils is more likely to occur in the United States due to greater water pressure, floods, typhoons and earthquakes.

Researchers estimated that shrinking and swelling soils cause approximately $2.3 billion in damage annually to small buildings and road surfaces in the United States. This amount of damage is twice the amount of damage incurred from floods, earthquakes and hurricanes. Researchers estimated that swelling soils cause approximately $7 billion in damage each year. According to researchers 60% of 250,000 newly constructed homes incur minor expansive soil damages and 10% incur significant expansive soil damage each year in the United States. Researchers noted that expansive soils caused $490,000 in damage to a building over a 6-year period. The estimated annual cost due to significant structural damages, such as cracked driveways, sidewalks and basement floors, heaving of roads and highway structures, condemnation of buildings; and disruption to pipelines and other utilities in Colorado, is $16 billion, according to AMEC.

Swelling pressure depends on the type of clay mineral, soil structure and fabric, cation exchange capacity, pH, cementation and organic matter. Any cohesive soil can involve clay minerals, but montmorillonite or bentonite clay minerals are more active regarding swelling-shrinkage. Swelling is calculated by swelling experiments with chemical and mineralogical analysis, soil indices and some empirical formulas from soil classifications. The shrinkage limit is determined from a laboratory test or approximate calculation recommended by Casagrande. Properties of clay improve with chemical additives such as cement, lime, lime-fly ash, cement-fly ash, calcium chloride and so on.

Structures transfer loads to the subsoil through their foundations. The imposed stress from the structure compresses the subsoil. This compression of soil mass leads to a decrease in the volume of the mass, which results in the settlement of the structure, and this should be kept within tolerable limits. Therefore, settlement (compression) should be estimated before construction. The settlement is defined as the compression of a soil layer due to the construction of foundations or other loads.

The compression is seen in deformation, relocation of soil particles and expulsion of water or air from void spaces. In general, the soil settlement under load falls into three categories: immediate or elastic settlement, which is caused by the elastic deformation of dry soil or moist and saturated soils without change in the moisture content; primary consolidation settlement, which is the result of a volume change in saturated cohesive soils because of the expulsion of water occupying void spaces; and secondary consolidation settlement is the volume change under a constant effective stress due to the plastic adjustment of soil fabrics. The consolidation settlement is seen when a structure is built on saturated clay or the water level is permanently lowered.

Simultaneously, consolidation settlement is seen under its own weight or the weight of soils that exists above the clay. The consolidation settlement of clay takes a long time, and the reason for this is the low hydraulic conductivity and slow drainage of clay. Settlement of the soil is determined by one-dimensional consolidation (odometer) and hydraulic consolidation (Rowe). In experiments, the vertical loads and void ratios are recorded. Afterwards, the relationship between the pressure and void ratio is determined from the measured data. These data are also useful in determining the consolidation coefficient. The consolidation coefficient is determined by the root of time method and the log-t method.

Shear strength behavior of clay

The shear strength of soils is one of the most important aspects of geotechnical engineering. The strength of the soil provides safety for geotechnical structures. The bearing strength, slope stability and bearing wall of the bases are influenced by the shear strength of the soils. Failure in the soils occurs in the form of shear. If the stresses in the soil exceed the shear strength, failure occurs. The shear failure of the soil depends on the interactions between the soil particles. These interactions are divided into friction strength and cohesion strength.

When the clay soils are subjected to shear, the volume change in the drainage shear depends on the environmental pressure, as well as the stress history of the soil. In addition, loading on clay soils does not allow water to escape from the pores, and thus, this creates excess water pressure. If the loading does not cause failure, the excess water pressure is dampened, consolidation occurs and volume change is observed. The long process of this volume change in the clays is due to very low hydraulic conductivity. Determination of the shear strength of the clay is performed by a direct shear test, triaxial compression test, vane test and standard penetration tests.

Physico-chemical and microstructure behavior of clay

For the determination of the physico-chemical and microstructural properties of clay soils, X-ray diffractometer (XRD) and scanning electron microscope (SEM) are commonly conducted. In addition, to determine the physico-chemical properties and structure of the soils, a pH test, electrical conductivity, cation exchange capacity, helium pycnometer, mercury intrusion porosimetry (MIP), surface area analysis (SSA), Brunauer-Emmett-Teller (BET) method or likewise, zeta potential and wavelength dispersive X-ray fluorescence test and Differential Thermal Analysis (DTA) are conducted. The pH value indicates the degree of H^+ or OH^- ions present. The change in pH affects the soil-water relations. Low pH indicates flocculation, and high pH indicates dispersion.

The electrical conductivity of clay is defined by its ion number and type. Cation exchange capacity is a measure of isomorph displacement capacity. Isomorph displacement is when other ions of equal or different valence to those of the ions are left. This change emerges from the unbalanced electrical charge for every change. To prevent this imbalance, the cations in the environment enter the edges of the clays and between the blocks.

X-ray diffractometer (XRD) analyses: The mineralogical composition of soils is critical due to its significant influence on soil behavior; the soils are affected at first degree, especially by physical, chemical and mechanical properties of clay and by the mineral content. In geotechnics, it is important to find the type of minerals present in clay, as well as their proportions to understand the mechanical behavior. The X-ray diffraction patterns of clay show a mineralogical composition of montmorillonite, anorthite, quartz, calcite and silica.

Fig. 3. The SEM images of typical clay for different magnification (a. 1000×, b. 10,000×, c. 35,000×).

Scanning electron microscope (SEM): The microstructure of soils, especially clays, is observed using a versatile, analytical and ultrahigh-resolution field-emission SEM. An SEM provides a high level of magnification. Soil specimens that are magnified up to 1,000,000 times enable the evaluation of differences in the surface by imaging the surface structures. The changes in the microstructural development of soils play a significant role in the behavior of soils. In particular, these parameters could lead to a better understanding of the engineering properties of compacted soils. The SEM images of typical clays are present in Fig. 3. Thus, flocculated and dispersed structures are observed in the soil samples.

Surface area analysis (SSA): The specific surface area is affected by grain-size distribution and the types and amounts of different clay minerals. Specific surface area is affected by the physical and chemical properties of soils.

3

PERMEABILITY AND SEEPAGE

INTRODUCTION

Permeability, as the name implies (ability to permeate), is a measure of how easily a fluid can flow through a porous medium. In geotechnical engineering, the porous medium is soils and the fluid is water at ambient temperature. Generally, coarser the soil grains, larger the voids and larger the permeability. Therefore, gravels are more permeable than silts. Hydraulic conductivity is another term used for permeability, often in environmental engineering literature.

DEFINITION OF PERMEABILITY

It is defined as the property of a porous material which permits the passage or seepage of water (or other fluids) through its interconnecting voids.

A material having continuous voids is called permeable. Gravels are highly permeable while stiff clay is the least permeable, and hence such a clay may be termed impermeable for all practical purpose.

Flow of water through soils is called seepage. Seepage takes place when there is difference in water levels on the two sides of the structure such as a dam or a sheet pile as shown in Fig. 1. Whenever there is seepage (e.g., beneath a concrete dam or a sheet pile), it is often necessary to estimate the quantity of the seepage, and permeability becomes the main parameter here.

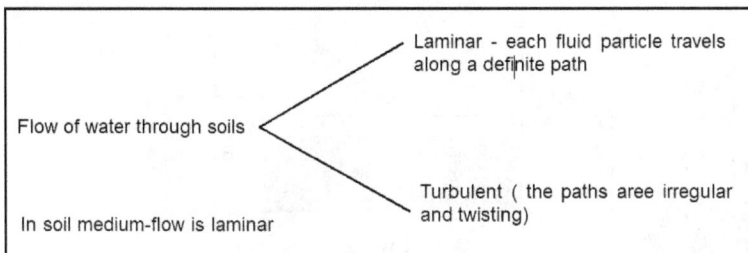

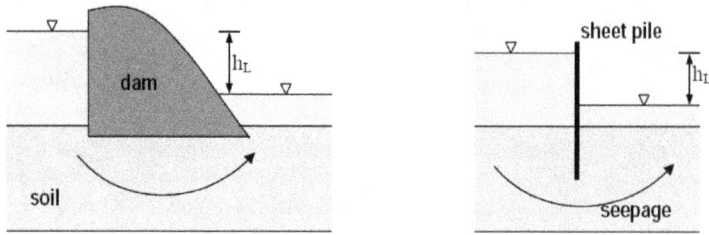

Fig. 1 Seepage beneath (a) a concrete dam (b) a sheet pile

Sheet piles are interlocking walls, made of steel, timber or concrete segments. They are used water front structures and cofferdams (temporary structure made of interlocking sheet piles, making up an impermeable wall surrounding an area, often for construction works.

The study of seepage of water through soil is important for the following engineering problems:

1. Determination of rate of settlement of a saturated compressible soil layer.
2. Calculation of seepage through the body of earth dams and stability of slopes for highways.
3. Calculation of uplift pressure under hydraulic structure and their safety against piping.
4. Groundwater flow towards well and drainage of soil.

DARCY'S LAW (1856) OF PERMEABILITY:

For laminar flow conditions in a saturated soil, the rate of the discharge per unit time is proportional to the hydraulic gradient.

$q = kiA$

$v = q/A = Ki$ (1)

Where

q = discharge per unit time

A = total cross-sectional area of soil mass, perpendicular to the direction of flow

i = hydraulic gradient

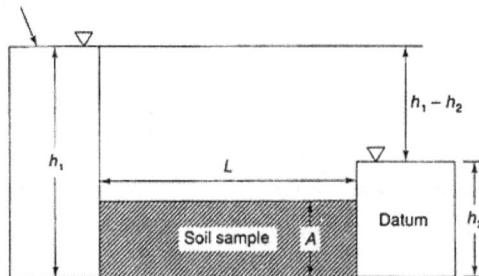

Fig. 2

k = Darcy's coefficient of permeability

v = velocity of flow or average discharge velocity

If a soil sample of length L and cross-sectional area A, is subjected to differential head of water $h_1 - h_2$, the hydraulic gradient i will be equal to $[(h_1 - h_2)/L]$ and we have q = k. $[(h_1 - h_2)/L]$.A.

When hydraulic gradient is unity, k is equal to V. Thus, the coefficient of permeability, or simply permeability is defined as the average velocity of flow that will occur through the total cross-sectional area of soil under unit hydraulic gradient. Dimensions are same as of velocity, cm/sec.

The coefficient of permeability depends on the particle size and various other factors. Some typical values of coefficient of permeability of different soils are given in Table 1.

Table 1. Typical values of permeability (k)

Soil Type	k is of the order of, in cm/dec
Gravel	10^1 to $1(10^0)$
Coarse sand	10^0 to 10^{-1}
Medium sand	10^{-1} to 10^{-1}
Fine sand	10^{-2} to 10^{-3}
Silty sand	10^{-3} to 10^{-4}
Delhi silt	7×10^{-5}

DISCHARGE VELOCITY AND SEEPAGE VELOCITY

The total cross-sectional area of the soil mass is composed of sectional area of solids and voids, and since flow cannot occur through the sectional areas of solids, the velocity of flow is merely an imaginary or superficial velocity.

The true and actual velocity with which water percolates through a soil is called the velocity of percolation or seepage velocity. It is the rate of discharge of percolating water per unit of net sectional area of voids perpendicular to the direction of flow.

VALIDITY OF DARCY'S LAW

In accordance with the Darcy's Law, the velocity of flow through soil mass is directly proportion to the hydraulic gradient for laminar flow condition only. It is expected that the flow to be always laminar in case of fine-grained soil deposits because of low permeability and hence low velocity of flow.

However, in case of sands and gravels flow will be laminar upto a certain value of velocity for each deposit and investigations have been carried out to find a limit for application of Darcy's law.

According to researchers, flow through sands will be laminar and Darcy's law is valid so long as Reynolds number expressed in the form is less than or equal to unity as shown below –

$$\frac{v \, D_a \, g_w}{\eta g} < 1 \qquad (2)$$

Where v = velocity of flow in cm/sec

D_a = size of particles (average) in cm.

It is found that the limiting value of Reynolds number taken as 1 is very approximate as its actual value can have wide variation depending partly on the characteristic size of particles used in the equation.

Factors affecting permeability are:

1. Grain size
2. Properties of pore fluid
3. Void ratio of the soil
4. Structural arrangement of the soil particle
5. Entrapped air and foreign matter
6. Adsorbed water in clayey soil

CAPILLARITY-PERMEABILITY TEST

The set-up for the test essentially consists of a transparent tube about 40 mm in diameter and 0.35 m to 0.5 m long in which dry soil sample is placed at desired density and water is allowed to flow from one end under a constant head, and the other end is exposed to atmosphere through air vent.

At any time interval t, after the commencement of the test, Let the capillary water travel through a distance x, from point P to Q. At point P, there is a pressure deficiency (i.e., a negative head) equal to h_c of water.

PERMEABILITY OF STRATIFIED SOIL DEPOSITS

In general, natural soil deposits are stratified. Each layer may be homogeneous and isotropic. When we consider flow through the entire deposit the average permeability of deposit will vary with the direction of flow relative to the bedding plane. The average permeability for flow in horizontal and vertical directions can be readily computed.

Average Permeability Parallel to Bedding Plane

Figure 3 shows several layers of soil with horizontal stratification. Let Z_1, Z_2, ...Z_n be the thickness of layers with permeabilities k_1, k_2, ... k_n.

For flow parallel to bedding plane the hydraulic gradient i will be same for all layers.

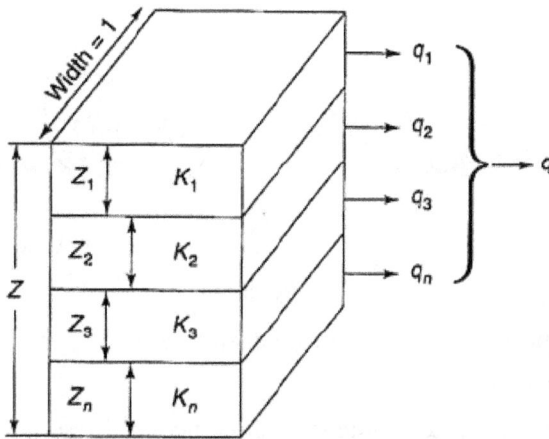

Fig 3 Flow parallel to bedding plane

Average Permeability Perpendicular to Bedding Plane

For flow in the vertical direction for the soil layers shown in Fig. 4.

In this case the velocity of flow, v will be same for all layers the total head loss will be sum of head losses in individual layers.

$$h = h_1 + h_2 + h_3 + \dots + h_n \tag{3}$$

SOIL PERMEABILITY

A soil mass is composed of small solid particles which we call the soil grains. These soil grains when depositing in a soil mass arranges themselves in a way that some amount of empty space is left between them. We call these empty spaces voids (Fig. 5).

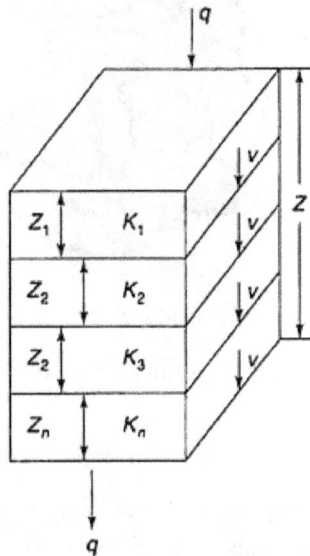

Fig. 4. Perpendicular to Bedding Plane

Soil Grains

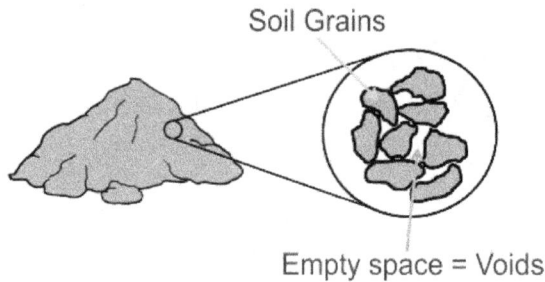

Empty space = Voids

Fig. 5

These voids or pores are interconnected and form a highly complex network of irregular tube like structure.

When water is subjected to a potential difference in the soil, it flows through these voids from high potential to low potential. The surface of the soil particles offers a resistance to the flow of water. The more irregular and narrower the voids, greater is the resistance posed to the water flowing while more open the voids, greater is the ease with which water flows through soil.

This property of the soil which permits the water or any liquid to flow through it through its voids is called permeability. It is the ease with which water can flow through the soils.

Gravel particles are large. Larger the soil grains, larger will be the volume of voids and better will be the connectivity of those pores, consequently large amount of water may flow through them easily and higher will be the flow of water, and that we say higher is the permeability of the soil.

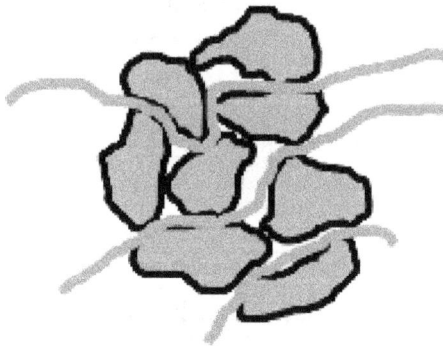

Fig. 6

A soil has different values of permeability for different liquids but we will focus only on soil's permeability with water.

Gravel soils are most permeable while clay soils are least permeable (Fig. 7).

Clay soils have high void ratio as they have large volume of voids because of their flocculated soil structure. These voids in the figure look very large but it is much zoomed view and because of clay particles being very small these voids are poorly connected to each other and do not form a regular tube or channel like structure. Hence even after having large amount of voids the clay soils are very less permeable.

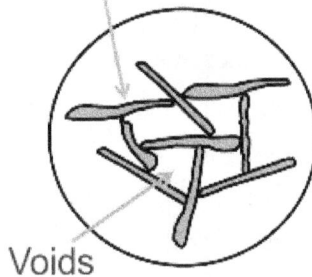

Fig. 7 Flocculated soil structure

When a soil has extremely low permeability it is termed as impervious soil.

The ability of soil to allow flow of water through it is called as *permeability of soil*. It is very important factor for the structures which are in contact with water. Flow of water in soil takes place through void spaces, which are interconnected. Water does not flow in a straight line, but in a winding path. However in soil mechanics flow is considered to be in a straight line at an effective velocity. The velocity of flow depends on size of pores.

IMPORTANCE OF PERMEABILITY OF SOIL

- ◉ Permeability influences the rate of settlement of a saturated soil under load.
- ◉ The stability of slopes and retaining structures can be greatly affected by the permeability involved.
- ◉ The design of earth dams is very much based upon the permeability of soil used.
- ◉ Filters made of soils are designed based upon their permeability.

Properties of Permeability of Soil

- ◉ Solving problems involving pumping seepage water from construction excavation.
- ◉ Estimating the quantity of underground seepage.
- ◉ Stability analysis of earth structures and earth retaining walls subjected to seepage forces.

FACTORS AFFECTING PERMEABILITY OF SOIL

- ◉ Grain size or Particle size

$K = CD_{10}^2$

The above equation is given by Alan Hazen. Permeability depends on shape and soil of soil particles.Permeability varies with square of particle size diameter.

- ◉ Void Ratio

If the presence of voids is more then the permeability is also more.

$$K = \frac{e^3}{1+3}$$

- ◉ Composition

For gravels, sand and silts presence of mica can decrease the *permeability of soil.*

For clay, water attracted between clay particles reduces the permeability.

- ◉ Structural Arrangement

Remolding of natural soil reduces permeability. If soil contains more rounded particles, the permeability is more.

- ◉ Stratification

When flow of water is parallel to strata, permeability will be more when compared with flow perpendicular to strata.

- ◉ Presence of foreign particles and entrapped air

This affects the *permeability* as it reduces void space and it blocks the inter-connectivity between the pores.

- ◉ Degree of saturation

If the soil is dry or partly saturated the permeability of soil is always less.

DARCY'S LAW

In 1856, french hydraulic engineer Henry Darcy published a report on the water supply of the city of Dijon in France. In that report, Darcy described the result of an experiment designed to study the flow of water through a porous medium. Darcy's experiment resulted in the formulation of mathematical law that describes fluid motion in porous media. Darcy's law states that the rate of fluid flow through porous medium is proportional to the potential energy gradient within that fluid. The constant of proportionality is the Darcy's *permeability of soil.* Darcy's permeability is a property of both porous medium and the fluid moving through the porous medium. In fact, Darcy's law is the empirical equivalent of the Navier-Strokes equation. Darcy's flow velocity for laminar flow is defined as the quantity of fluid flow along the hydraulic gradient per unit cross sectional area. Velocity of flow through a porous media is directly proportional to the hydraulic gradient responsible for flow.

$$V = Ki = K \frac{\Delta h}{L}$$

Here,

- ◉ V = discharge velocity or superficial velocity
- ◉ k = coefficient of permeability or hydraulic conductivity

- ⊙ i = hydraulic gradient
- ⊙ Δ = fall in total head
- ⊙ L = length of soil specimen

Assumptions of Darcy's Law

- ⊙ Soil is fully saturated.
- ⊙ Temperature during testing is 27°C.
- ⊙ Flow through soil is laminar
- ⊙ Entire cross sectional area is available for flow
- ⊙ Flow is continuous and steady.

VALIDITY OF DARCY'S LAW

Darcy's law is valid only for slow and viscous flow, fortunately most groundwater flow cases fall in this category. Typically any flow with a Reynolds number less than 1 is clearly laminar and it would be valid to apply Darcy's law. Experimental tests have shown that flow regimes with values of Reynolds number upto 10 may still be Darcine.

SEEPAGE

Let's have a look at the concrete dam and sheet pile in Fig. 1, where seepage takes place through the sub soil, due to head difference between up stream and down stream water levels. If we know the permeability of the soil, how do we compute the discharge through the soil? How do we compute the pore water pressures at various locations in the flow region or assess the uplift loading on the bottom of the concrete dam? Is there any problem with hydraulic gradient being too high within the soil? To address all these, let's look at some fundamentals in flow.

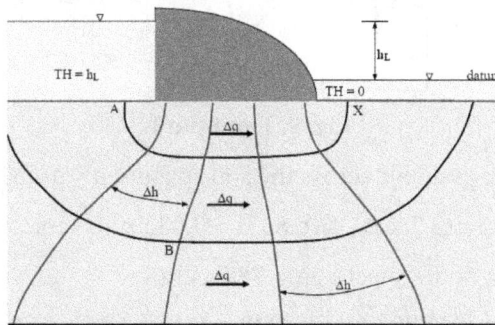

Fig. 8 Streamlines and equipotential lines

In the flow beneath the concrete dam shown in Fig. 8, we will assume that the concrete dam is impervious, and an impervious stratum such as bedrock or stiff clay underlies the soil.

Let's select the datum as the downstream water level. That makes the total head within the downstream and upstream water 0 and hL respectively. There is total head loss of hL along each streamline originating upstream and ending up downstream. This loss in head or energy is used against overcoming the resistance to flow provided by the soil. Here, streamline or flow line, is the path of a water molecule in the flow region. There are thousands of streamlines in the flow region. The flow passage between two adjacent streamlines is known as flow channel. In a flow net, we only draw streamlines such that the discharge is the same (= Δq) through each of the flow channels. Let's say there are Nf flow channels in the flow net.

An equipotential line is a contour of constant total head. The blue lines shown in the figure are all equipotential lines, where the total head is constant along each of them. In a flow net, such as the one shown in Fig. 8, the equipotential lines are drawn such that the total head difference between two adjacent ones is the same (= Δh) throughout the flow region. If there are Nd equipotential drops in a flow net, Δh = hL/Nd. In Fig. 8, Δh = 0.2 hL.

Example 1

Flow takes place through a 100 mm diameter and 275 mm long soil sample, from top to bottom, as shown in the Fig 9. The manometers are 120 mm apart, and the water level difference within the two manometers is 100 mm at steady state. If the permeability of the soil is 3.7×10^{-4} cm/s, what is the flow rate?

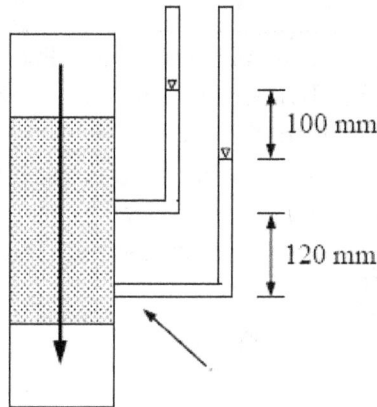

Fig. 9. Example 1

Solution: Hydraulic gradient across the soil specimen = 100/120 = 0.833

Velocity of flow = k i =(3.7 x 10-4)(0.833) = 3.082 x 10^{-4} cm/s

Cross sectional area of the specimen = 78.54 cm²

Flow rate = (3.082 x 10^{-4})(78.54) = 0.0242 cm³ /s = 1.45 cm³ /min

Example 2

A long horizontal drain at 3 m depth collects the ground water in a low-lying area. The free water table coincides with the ground level and the flownet for the ground water flow is shown in Fig. 1. The 6 m thick sandy clay bed is underlain by an impervious stratum. Permeability of the sandy clay is 6.2×10^{-5} cm/s.

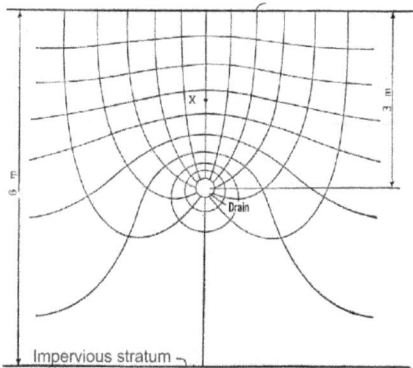

Fig. 10. Flownet around the Horizontal Drain

a. Find the discharge through the drain in m³/day, per metre length of drain.

b. Find the pore water pressure at X, 1.5 m into the soil, directly above the drain.

c. Estimate the velocity of flow at X.

Solution:

Let's take datum at the drain level. This makes the total head at the top of the ground (and water table at the surface) 3.0 m. Head loss from the surface to drain = 3.0 m.

Let's assume water is at atmospheric pressure in the drain and at the surface (ground level).

a. Let's consider only half of the flow region.

$h_L = 3.0$ m; $N_f = 6$; and $N_d = 9$

$$Q = kh_L \frac{N_f}{N_d}$$

$= (6.2 \times 10^{-7}$ m/s$)(3$m$)(2 \times 6/9)(24 \times 3600) = 0.214$m³/day per m

b. Change in total head between two equipotential drops "h = 3/9 = 0.333 m.

Total head at X = 3.0 –3.4 Dh =1.867 m

Elevation head at X = 1.5 m\/

Pressure head at X = 1.867 – 1.500 = 0.367 m

Pore water pressure at X = (0.367)(9.81) = 3.60 kPa

c. The vertical length of the curvilinear square at X is measured as 0.4 m

Hydraulic gradient at X = 0.333/0.4 = 0.83

Velocity = k i = 6.2 × 10-5 × 0.83 = 5.15 × 10-5 cm/s.

4

COMPACTION

INTRODUCTION

In geotechnical engineering, soil compaction is the process in which a stress applied to a soil causes densification as air is displaced from the pores between the soil grains. When stress is applied that causes densification due to water (or other liquid) being displaced from between the soil grains, then consolidation, not compaction, has occurred. Normally, compaction is the result of heavy machinery compressing the soil, but it can also occur due to the passage of (e.g.) animal feet.

In soil science and agronomy, soil compaction is usually a combination of both engineering compaction and consolidation, so may occur due to a lack of water in the soil, the applied stress being internal suction due to water evaporation as well as due to passage of animal feet. Affected soils become less able to absorb rainfall, thus increasing runoffand erosion. Plants have difficulty in compacted soil because the mineral grains are pressed together, leaving little space for air and water, which are essential for root growth. Burrowing animals also find it a hostile environment, because the denser soil is more difficult to penetrate. The ability of a soil to recover from this type of compaction depends on climate, mineralogy and fauna.

Soils with high shrink-swell capacity, such as vertisols, recover quickly from compaction where moisture conditions are variable (dry spells shrink the soil, causing it to crack). But clays which do not crack as they dry cannot recover from compaction on their own unless they host ground-dwelling animals such as earthworms — the Cecil soil series is an example.

Soil is used as a construction material for constructing embankments and subgrades. Embankments are constructed to raise the ground level above the existing level up to the formation level to support buildings, roads, or railways or other structures and also to retain water as in the case of earth dams or canal banks. Subgrades are constructed to provide support to the roads as a base to withstand the traffic loads.

The stability and durability of the embankments and subgrades depends on the improvement of shear strength of the soil as well as the restriction of settlements or deformation within permissible limits. The strength and deformation characteristics of embankments and subgrades depend directly on the density of the soil.

The higher the density, the higher is the strength and lesser is the settlement under loads. The seepage of water through embankments and subgrades acts to reduce the strength due to its erosive nature. Higher density will result in reduced permeability of the soil, thereby reducing the seepage of water.

Densification of soils during construction of embankments and subgrades is achieved by compaction. Higher density of embankments and subgrades is achieved by compacting the soil by rollers usually in layers, known as lifts.

DEFINITION OF COMPACTION

Compaction is the artificial and mechanical process of decreasing the volume of the soil rapidly by the expulsion of air voids in the soil resulting in the increase in density.

Densification of soil also occurs naturally due to consolidation of foundation soils by expulsion of pore water due to loads from the structure. This is a rather long-term process compared to compaction.

CAUSES OF SOIL COMPACTION

There are several forces, natural and man-induced, that compact a soil. These can be great, such as from a tractor, combine or tillage implement, or it can come from something as small as a raindrop. Listed below are types of soil compaction and their causes.

Raindrop Impact

This is certainly a natural cause of compaction, and we see it as a soil crust (usually less than ½ inch thick at the soil surface) that may prevent seedling emergence. Rotary hoeing can often alleviate this problem. Drops of rain cause splash erosion, which disturbs the top layer of soil particles and causes formation of a thin surface crust that blocks water from reaching plant roots.

Tillage Operations

Tilling the soil accelerates breakdown of organic materials that inhibit compaction. It can also damage soil structure, the arrangement of mineral particles in relation to pore space, especially if soil is tilled when it is wet. Over the years, repeated tillage orients all of the soil particles in the same direction, causing a layer of compacted soil (a plow pan) to form directly beneath the area being tilled. Plow pans are mainly a problem on farm fields where the soil is consistently tilled at the same depth. Continuous mouldboard plowing or disking at the same depth will cause serious tillage pans (compacted layers) just below the depth of tillage in some soils. This tillage pan is generally relatively thin (1-2 inches thick), may not have a significant effect on crop production, and can be alleviated by varying depth of tillage over time or by special tillage operations.

Wheel Traffic

This is without doubt the major cause of soil compaction. With increasing farm size, the window of time in which to get these operations done in a timely manner is often limited. The weight of tractors has increased from less than 3 tons in the 1940's to approximately 20 tons today for the big four-wheel-drive units. This is of special concern because spring planting is often done before the soil is dry enough to support the heavy planting equipment. Heavy farm machinery can create persistent subsoil compaction. An axle load of 10 Mg can cause compaction to penetrate to a depth of 50 cm, and with still higher loads, compaction can reach 1 m deep.

Loads as high as 30 Mg per axle are used in many countries. Consequently, soil compaction resulting from farm machinery has become a major concern in agriculture worldwide. Soil is also compacted during building construction, from repeated use of riding lawn mowers, or from off-road parking of automobiles and recreational vehicles. Pedestrian pathways across garden beds and turf areas are also significant contributors to compaction.

Intensity of trafficking (number of passes) plays an important role in soil compaction because deformations can increase with the number of passes. Experimental findings have shown that all soil parameters become less favourable after the passage of a tractor and that a number of passes on the same tramlines of a light tractor, can do as much or even greater damage than a heavier tractor with few passes.

Minimal Crop Rotation

The trend towards a limited crop rotation has had two effects: (1) Limiting different rooting systems and their beneficial effects on breaking subsoil compaction, and (2) Increased potential for compaction early in the cropping season, due to more tillage activity and field traffic.

Natural Processes

Soils with high clay content—typical of wetlands and river bottoms—can become compacted due to natural processes. Because individual clay particles are so small, they are more susceptible to being pressed together tightly.

Pasture Grazing

There is a tendency for the soil to become compacted after continuous grazing of animals on the field. Researchers reported that regardless of the type of grazing system soil bulk density is increased at field moist condition, and that this effect was most pronounced at soil depths less than 10 cm. He further revealed that there was an increase in penetration resistance and a general reduction in soil water up to a depth of 15 cm. Researchers stated that poor grazing management is one of the major causes of soil compaction in agricultural land. According to him livestock traffic develops soil compaction due to repeated pressure in the area due to poor grazing management. Although livestock can break the upper layer of the soil due to hoof action, deep compaction layers develop overtime if left untreated. Soils that are higher in clay content are more susceptible to hoof compaction than sandier soils. He further reiterated that aeration can decrease soil compaction and allow for greater plant root development.

THE DIFFERENCE BETWEEN COMPACTION AND CONSOLIDATION

The difference between compaction and consolidation is given below:

Compaction

1. Artificial process caused by mechanical means such as rollers.
2. Decrease in volume and increase in the density of soil occurs by expulsion of air from the voids.
3. Compaction occurs in partially saturated soils.
4. Compaction is completed within minutes-and hence is a short-term process.
5. Compaction is effective in well-graded soils containing gravel and sand, and to a less extent in silts and clays.
6. Compaction is caused by short-term dynamic load, which are removed after compaction.

Consolidation

1. Natural process caused by stresses due to foundations or superstructures.
2. Decrease in volume and increase in the density occurs by expulsion of pore water from the voids.
3. Consolidation takes place in fully saturated soils.
4. Consolidation takes several months to years and hence is a long-term process.
5. Consolidation although in principle occurs in all soils but is significant for clayey soils from engineering point of view due to consequent long-term settlements.
6. Consolidation is caused by long-term static loads, which continue to exist after the completion of consolidation.

PRINCIPLE OF COMPACTION

The principle of compaction was developed by R. R. Proctor in 1933 during construction of earth dams in California. The objective of compaction is to achieve maximum possible dry density of the compacted soil. The water content used for compaction controls the dry density achieved. Figure 1 shows the variation of the dry density with water content.

At low water content, the soil is stiff and the particles offer resistance to come closer, resulting in low dry density. As the water content is increased, water forms a lubricating film around particles causing them to be compacted to a closer state of contact resulting in higher dry density. The dry density increases with increase in the water content until maximum dry density (MDD) is reached.

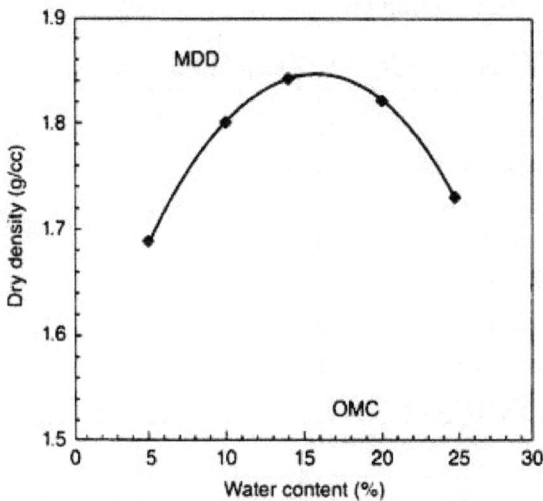

Fig. 1. Effect of water content on dry density

At this stage, the soil particles come to the closest possible state of contact. On increase of water content beyond optimum moisture content (OMC), the volume of soil does not decrease further by compaction and water starts to occupy additional space causing an increase in the volume of voids and the total volume, and resulting in a decrease in dry density.

The water content at which the dry density is maximum after compaction is known as optimum moisture content or optimum water content. In general, water equal to OMC is added in the field for effective compaction, except in some specific cases. Compactive effort or compaction energy also controls the effectiveness of compaction. Higher the compactive effort, higher will be the dry density achieved for the same soil.

The type of soil and its gradation and plasticity characteristics also influence the degree of compaction achieved. Coarse-grained soils can be compacted to a higher dry density than fine-grained soils. Cohesionless soils can be similarly compacted to a higher dry density than cohesive soils. A well-graded soil is compacted more effectively as compared to a poorly graded soil. Addition of fines to a coarse-grained soil, by an amount just required to fill the existing voids, greatly enhances the dry density.

For the compaction of a given soil, the sample of soil is compacted in the laboratory applying standard compaction energy at different water contents. The dry density of the compacted soil at each of the water content is determined and a graph is plotted with the water content on the x-axis and the dry density on the y-axis.

The water content corresponding to maximum dry density is determined, which gives optimum water content. For the compaction of soil in the field, water equal to OMC, or less (dry of OMC) or more (wet of OMC) water is used depending on the objective of compaction and type of construction. Same compaction energy per unit volume of soil, as used in the laboratory compaction test, is used to compact the soil in the field.

THE OBJECTIVES OF COMPACTION

The following are some of the objectives of compaction:

   i. Increase the shear strength of soil.
   ii. Decrease the undesirable settlement of structures.
   iii. Control undesirable volume change.
   iv. Decrease permeability of soil.
   v. Increase the bearing capacity of foundations.
   vi. Increase the stability of slopes.

EFFECT OF COMPACTION ON ENGINEERING PROPERTIES OF THE SOIL

Compaction improves the strength and deformation characteristics of the soil, improving their stability and durability. Lambe (1958) found that the properties of soil after compaction depend on the soil structure, which, in turn, is influenced by the type of soil, amount of water relative to OMC, and the compaction energy applied.

The effect of compaction is discussed on the following soil properties:

Soil Structure

Soil compacted at the water content less than OMC (dry of optimum) will have flocculent structure with edge-to-face particle arrangement, irrespective of method of compaction. The structure of soils compacted at water content greater than OMC (wet of optimum) depends on the magnitude of the shear strain. Soils compacted wet of optimum, which undergo relatively small shear strain during compaction, will have flocculent structure. Soils compacted wet of optimum, which undergo large shear strains during compaction, usually have a dispersed structure with face-to-face (oriented) particle arrangement.

The degree of orientation of soil particles increases gradually with increase in water content and the soil still possesses a flocculated structure up to the OMC. The orientation of particles increases more rapidly with increase in water content for soils compacted wet of optimum.

Increase of compaction energy increases the orientation of soil particles even at the same water content.

Shear Strength

Soils compacted dry of optimum have more shear strength than those compacted wet of optimum. The cohesion and friction angle are both higher for soils compacted dry of optimum. Thus, the Mohr-Coulomb strength envelope is steeper for soils compacted dry of optimum and also lies above that of soils compacted wet of optimum. However, the difference in shear strength of soils compacted dry and wet of optimum decreases when the compacted soils are fully saturated.

It may be noted that soils with a flocculent structure possess more shear strength. This is because the attractive forces are predominant in flocculent structure and also because the soil offers higher resistance to deformation due to particle interference in edge-to-face particle arrangement existing in flocculent structure.

On the other hand, repulsive forces are predominant in soils with dispersive structure resulting in lower shear strength. The particle interference and hence the resistance to deformation is also less in dispersed structure, which has relatively oriented particle arrangement.

Saturation of compacted soils increases the repulsive forces, causing a decrease in shear strength.

Pore Water Pressure

As the water content is less for soils compacted dry of optimum, there is zero or negligible pore water pressure (due to discrete and local pockets of saturation). Soils compacted wet of optimum show higher pore water pressure, which reduces the effective stress and frictional component of shear strength.

Stress-Strain Relationship

Soils compacted dry of optimum possess a steeper stress-strain relationship compared to those compacted wet of optimum. Consequently, the deformation and settlement are less for soils compacted dry of optimum, and show relatively sudden and brittle failure. Soils compacted wet of optimum show large strains and settlements and the failure is gradual and plastic.

Compressibility

Soils compacted dry of optimum are less compressible due to their flocculent structure and greater particle interference and resistance to deformation. Soils compacted wet of optimum are initially less compressible at low stresses due to their dispersed structure and predominance of repulsive forces.

However, when the stresses are increased further to overcome the repulsive forces, such soils show high compressibility resulting in large deformation. The face-to-face particle arrangement in dispersed structure of such soils also offers less resistance to deformation and increases the compression.

Shrinkage

Shrinkage is the decrease in the volume of soil due to the evaporation of water. Soil compacted dry of optimum undergoes less shrinkage due to random particle arrangement and particle interference that offers more resistance to deformation. Shrinkage is more for soils compacted wet of optimum due to dispersed structure and lesser particle interference and resistance to deformation.

Swelling

A clay soil compacted dry of optimum has more water deficiency and large void ratio and hence imbibes more water resulting in larger swelling, compared to the soil at the same dry density compacted wet of optimum.

Permeability

Soils compacted at low water content possess low dry density and large void ratio and hence are more permeable. With increase in water content dry of optimum, the dry density increases and void ratio decreases causing a decrease in permeability.

Thus, permeability of soils compacted dry of optimum decreases with increase in water content. Permeability is minimum at or slightly above the OMC. With further increase in water content, permeability slightly increases due to decrease in dry density. However, permeability of soils compacted wet of optimum is always much less than those compacted dry of optimum.

FACTORS AFFECTING COMPACTION

The MDD achievable by the compaction depends on the following factors:

1. Water content.
2. Type of soil and its gradation.
3. Gradation of Soil
4. Compaction energy.
5. Method of compaction.

Effect of Water Content

Increase of water content used for compaction increases the dry density initially until the dry density reaches its maximum. After reaching MDD, further increase in the water content decreases the dry density.

Type of Soil

The type of soil used for compaction primarily decides MDD achievable by the compaction.

Coarse-grained soils can be compacted to a higher dry density than fine-grained soils. Cohesive soils usually have high air voids content. The void ratio of cohesive soils increases with increase in plasticity index. Thus, coarse-grained soils have higher MDD and lower OMC than fine-grained soil. The MDD decreases and OMC increases for low plastic silt, high plastic silt, and high plastic clay.

Gradation of Soil

For a given soil, a well-graded soil has higher MDD and lower OMC then a poorly graded soil. This is because a well-graded soil contains particles of all sizes and the finer size particles fill the void space between the coarser particles resulting in lower air voids and higher MDD.

Addition of small amount of fines to a coarse-grained soil increases its MDD for the same reason. However, when the amount of fines added is more than that needed to fill the voids of coarse-grained soil, the MDD again decreases.

Compaction Energy

The compaction energy applied to the soil during compaction has a significant influence on the MDD. In general, the higher the compaction energy or compactive effort, the higher will be the MDD and lower will be the OMC. This is the reason why the subgrades of airfield pavements are compacted using heavy compaction.

The increase in dry density due to the increase in compactive effort is more at water content less than OMC (dry of optimum) than that on the wet of optimum.

It may be noted that the increase in compactive effort does not go on increasing the MDD indefinitely. When compactive effort is increased in equal increment, the increment in MDD becomes smaller and smaller with each increment of compactive effort. Finally, a stage is reached where further increase of compactive effort does not bring any significant increase in MDD.

Care should be taken to see that the compactive effort does not cause a stress on the soil particles beyond their crushing strength, in which case the higher compactive effort crushes the individual particles, causing a reduction in MDD in some soils.

The line joining the peak points of different compaction curves of the soil compacted with different compactive effort is known as line of optimums and is roughly parallel to the ZAVL.

METHOD OF COMPACTION

Compaction of soils in the field can be done by a variety of compaction equipment.

The following are the different actions or effects of various compaction equipment on soils:

1. Static compaction – smooth wheel rollers.
2. Kneading compaction – sheep's foot rollers.
3. Vibration compaction – vibratory rollers.
4. Tamping – tampers.

EFFECTS OF SOIL COMPACTION ON GROWTH AND YIELDS OF CROPS

It was reported that the ability of plant roots to penetrate soil is restricted as soil strength increases and ceases entirely at 2.5 MPa. Researchers reported that as cone index approaches 2. 0 MPa and moves above this value, root growth has been shown to be restricted to varying degrees.

Hence 2 MPa has been considered as a measure in the determination of soil hard pan layer. Researchers further revealed that critical limit of penetration resistance restraining root distribution is within 40-50 cm soil depth and that subsoiling can reduce and provide increased rooting depth. Researchers explained that hydrostatic pressure (turgor) within the elongating region of the root provides the force necessary to push the root cap and meristematic region through the resisting soil.

If the hydrostatic pressure is not sufficient to overcome wall resistance and soil impedance, elongation of that particular root tip ceases. Plant roots constitute a major source of soil organic matter when decomposed; and while growing, are capable of both creating and stabilizing useful structural features.

Researchers opined that excessive soil compaction impedes root growth and therefore limits the amount of soil explored by roots which in turn can decrease the plant's ability to take up nutrients and water, from the stand point of production. Plants grown in compacted soils have shown a smaller number of lateral roots than plants grown under controlled condition.

Plants grown in more compacted soils showed smaller ratios of fresh to dry mass. Soil compaction can have adverse effect upon crops by - increasing the mechanical impedance to the growth of roots; altering the extent and configuration of the pore space and aggravating root diseases. Plants grown in compacted soils have shown a smaller number of lateral roots than plants grown under controlled condition. Plants grown in more compacted soils showed smaller ratios of fresh to dry mass.

Soil compaction can have adverse effect upon crops by – increasing the mechanical impedance to the growth of roots; altering the extent and configuration of the pore space and aggravating root diseases. Researchers reported the effect of annual compaction (1987-1989) of 9 and 18 Mg axle loads, and subsoiling for a com/soybean rotation on a Hoytville silty clay loam (very poorly drained) soil. The effect on soil physical properties was also examined.

It was revealed that the 9 Mg and 18 Mg loads significantly reduced yields through 1992 and 1994, respectively. According to the researchers, subsoiling generally improved yields of all treatments including control. Researchers revealed that soil compaction caused yield reductions in cotton. Their study investigated the effect of soil compaction on canopy spectral reflectance, soil electrical conductivity (EC), and cotton yield.

Field experiments were conducted during 2003 2005 using a completely randomized block design with four soil compaction treatments. The treatments were no subsoiling (control); subsoiled, disked, and bedded (conventional); subsoiled and compacted (compaction I); and compacted with no subsoiling (compaction II). These results verified that compaction affected canopy reflectance and reduced cotton yields. The practical implications of the outcome of this study are the potential use of EC and canopy reflectance to infer crop yield and extent of soil compaction.

Triticale and maize, with different structure of the root system and type of photosynthesis were examined to know changes in shoot physiology and root architecture in response to varying degree of soil compaction. He reiterated that the effects of different levels of soil compaction (1.30, 1.47 and 1.58 Mg m^{-3}) on a shoot and root dry matter, leaf number and area, number and length of seminal, seminal adventitious, nodal and lateral roots, leaf water potential (ψ), maximum quantum yield of PS II (Fv/Fm) and gas exchange were studied in the root-box.

Severe soil compaction treatments decreased leaf number, leaf area and dry matter of shoots and roots, while increasing shoot-to-root dry matter ratio. In addition, high level of soil compaction strongly affected the length of seminal and seminal adventitious roots, and the number and length of lateral roots developed on the seminal root. Along with the restriction of root growth, significant influences were observed in ψ, Fv/Fm and gas exchange.

High soil compaction treatments resulted in decreased ψ, Fv/Fm, and photosynthetic rate, transpiration rate and stomatal conductance for both triticale and maize. Maize whose root growth was more heavily restricted by the soil compaction compared to triticale showed greater damages in physiological characteristics in leaves, while the impact on triticale was relatively small. The results indicated that damages in photosynthesis, water relation and shoot growth by soil compaction would be closely related to sensitivity of root systems architecture to high mechanical impedance of soil.

PREVENTION AND ALLEVIATION OF SOIL COMPACTION

One of the major challenges that farmers face in the field is the management of the soil compaction which occurs in their lands. This is a serious issue as it has a direct impact on the potential of your crop.

When preparing the soil for planting we need to have a clear goal in mind when it comes to compaction:

⦿ We should aim to limit its occurrence;

⦿ We must try to alleviate compaction which has already occurred; and

⦿ Ultimately we should strive to prevent compaction as best as we can.

When preparing to plant a crop we should have one primary consideration – the soil. The soil is the factory within which our produce will grow and develop. It is the foundation of the crop. If the foundation of your home is poorly constructed your walls will crumble in time. This is the same in our soils, if we do not look after the soil our potential income from the crops planted will also crumble in time.

For a seed to germinate and flourish it needs to have the correct environment in which to do so. So what is a good seed bed? To answer this question one should first consider your management practices on your farm. For example; do you practice no-tillage or do you implement conventional tillage? For the sake of this article let us use conventional tillage as an example as it is in conventional tillage practices where farmers especially battle with the challenge of soil compaction.

A good seed bed is made up of loosely aggregated soil fragments which are mostly free of soil clods and organic material. Soil should be loose for a depth of up to one metre. This will allow for good root penetration as well as moisture absorption into the soil profile. A goodseed bed is free of weeds as well as rocks. If your field is full of rocks then your planter will not penetrate effectively which will cause an uneven emergence of your crop.

A good seed bed is also level, which means there are no uneven ridges and gullies running through the field. Ridges and gullies will cause problems at planting time and will also have an adverse effect when it comes to soil erosion caused by run-off.

If soil is compacted we will not have a good seed bed.

⦿ When soil is compacted we don't have a loose aggregate in the soil.

⦿ We lose access to the entire soil profile.

⦿ We suffer considerable run-off which leads to erosion.

⦿ Consequently there is less absorption of moisture into the soil.

⦿ Planters will battle to penetrate.

⦿ Seeds will struggle to germinate and emerge.

⦿ We allow for the establishment of pioneer weed species which take advantage of the tough conditions.

To understand how to prevent and alleviate soil compaction we first need to understand what causes the compaction in the first place.

There are a number of factors which contribute to compaction. Firstly, the machinery which operates in your fields is heavy. Tractors, combines, trailers and trucks all compact the soil on which they drive. Moving wheels bind soil particles creating a hard surface. Another factor which causes compaction is animal traffic in your fields.

Just like the cattle paths which occur naturally on grazing land, so too when animals are let into the lands to graze crop residues their hooves cause compaction of soil. Soil which is left fallow will also compact over time due to natural weathering. Rainfall will wet and bind the soil which later gets baked by the sun causing a hard crust on the surface.

Good management practices

All these factors can be prevented with good management practices.

You can implement traffic control in your lands by driving on the same tracks whenever your machinery is working in the fields. This is becoming easier with the technology which is available to us today such us GPS technology and automatic steering equipment fitted into tractors. You can limit vehicle access into the lands by only allowing trucks and trailers to be loaded on the ends of the fields instead of inside the fields.

Compaction caused by animals can be prevented by carting the residues to the animals instead of allowing them to graze in the lands. Residues can be baled or captured directly from the combine and fed elsewhere. Natural weathering can't necessarily be prevented but it can be alleviated by the same methods used to alleviate compaction caused by the above mentioned factors.

To alleviate soil compaction in a conventional tillage system you should firstly start by doing a deep ripping of the soil. The ripper tines must be sharp and the best results are achieved when using a heavy implement operated by a strong tractor.

You want the ripper to penetrate deeply, this will also assist in ripping through old plough pans which is sub soil compaction, this occurs at the level under the soil where the implements work up to. This is caused by years of continuous soil tillage. Once the soil has been ripped, you can then either disc or plough the soil. After this, one should do a final soil working of harrowing. This will break the clods up causing the desirable loose soil aggregate for a good seed bed. Once this foundation is laid, one can now work on the building blocks of planting a good crop which will bare good yields for this season and the seasons to come.

SOIL COLOUR

Soil colour indicates many soil features. A change in soil colour from the adjacent soils indicates a difference in the soil's mineral origin (parent material) or in the soil development. Soil colour varies among different kinds as well as within the soil profile of the same kind of soil. It is an important soil properties through which its description and classification can be made.

Kinds of soil colour

Soil colour is inherited from its parent material and that is referred to as lithochromic, e.g. red soils developed from red sandstone. Besides soil colour also develops during soil formation through different soil forming processes and that is referred to as acquired or pedochromic colour, e.g. red soils developed from granite or schist.

Factors affecting soil colour

There are various factors or soil constituents that influence the soil colour which are as follows:

- ◉ Organic matter: soils containing high amount of organic matter show the colour variation from black to dark brown.

- ◉ Iron compounds: soil containing higher amount of iron compounds generally impart red, brown and yellow tinge colour.

- ◉ Silica, lime and other salts: Sometimes soils contain either large amounts of silica and lime or both.

- ◉ Due to presence of such materials in the soil the colour of the soil appears like white or light coloured.

- ◉ Mixture of organic matter and iron oxides: Very often soils contain a certain amount of organic matter and iron oxides. As a result of their existence in soil, the most common soil colour is found and known as brown.

- ◉ Alternate wetting and drying condition: During monsoon period due to heavy rain the reduction of soil occurs and during dry period the oxidation of soil also takes place.due to development of such alternating oxidation and reduction condition,the colour of soil in different horizons of the soil profile is variegated or mottled. This mottled colour is due to residual products of this process especially iron and manganese compounds.

- ◉ Oxidation-reduction conditions: when soils are waterlogged for a longer period, the permanent reduced condition will develop. The presence of ferrous compounds resulting from the reducing condition in waterlogged soils impart bluish and greenish colour.

Therefore, it may be concluded that soil colour indirectly indicative of many other important soil properties besides soil colour directly modify the soil temperature e.g.dark coloured soils absorb more heat than light coloured soils.

Determination of soil colour

The soil colours are best determined by the comparison with the Munsell colour.

This colour chart is commonly used for this purpose.the colour of the soil is a result of the light reflected from the soil. Soil colour rotation is divided into three parts:

Hue - it denotes the dominant spectral colour (red,yellow ,blue and green).

Value - it denotes the lightness or darkness of a colour (the amount of reflected light).

Chroma - it represents the purity of the colour (strength of the colour).

The Munsell colour notations are systematic numerical and letter designations of each of these three variables (Hue, Value and Chroma).For example ,the numerical notation 2.5 YR6/6 suggests a hue of 2.5 YR, value of 5 and chroma of 6. The equivalent or parallel soil colour name for this Munsell notation is 'red'.

SOIL WATER

Water, an excellent solvent for most of the plant nutrients,is a primary requisite for plant growth,

Importance of Soil Water

- Soil water serves as a solvent and carrier of food nutrients for plant growth
- Yield of crop is more often determined by the amount of water available rather than the deficiency of other food nutrients
- Soil water acts as a nutrient itself
- Soil water regulates soil temperature
- Soil forming processes and weathering depend on water
- Microorganisms require water for their metabolic activities
- Soil water helps in chemical and biological activities of soil
- It is a principal constituent of the growing plant
- Water is essential for photosynthesis

Water serves four functions in plants:

- it is the major constituents of plant protoplasm(85-95%)
- it is essential for photosynthesis and conversion of strarches to sugars
- it is the solvent in which nutrients move into and through plant parts to capture sunlight.
- In fact, the soil water is a great regulator of physical, chemical and biological activities in the soil.

Plants absorb some water through leaf stomata (openings), but most of the water used by plants is absorbed by the roots from the soil. For optimum water used, it is vital to know how water moves into and through the soil, how the soil stores water, how the plant absorbs it, how nutrients are lost from the soil by percolation, and how to measure soil water content and losses.

Soil also serves as a regulated reservoir for water because it receives precipitation and irrigation water.

A representative cultivated loam soil contains approximately 50% solid particles (sand, silt, clay and organic matter), 25% air and the rest 25%mater.only half of this water is available to plants because of the mechanics of water storage in the soil.

Structure of water

Water can participate in a series of reactions occurring in soils and plants, only because of its structural behavior. Water is simple compound, its individual molecules containing one oxygen atom and two much smaller hydrogen atoms.

The elements are bonded together covalently, each hydrogen or proton sharing its single electron with the oxygen. Instead of the atoms being arranged linearly (H-OH) the hydrogen atoms are attached to the oxygen as a v shaped.

Factors Affecting Soil Water

1. **Texture:** Finer the texture, more is the pore space and also surface area, greater is the retention of water.
2. **Structure:** Well-aggregated porous structure favors better porosity, which in turn enhance water retention.
3. **Organic matter:** Higher the organic matter more is the water retention in the soil.
4. **Density of soil:** Higher the density of soil, lower is the moisture content.
5. **Temperature:** Cooler the temperature, higher is the moisture retention.
6. **Salt content:** More the salt content in the soil less is the water available to the plant.
7. **Depth of soil:** More the depth of soil more is the water available to the plant.
8. **Type of clay:** The 2:1 type of day increases the water retention in the soil.

CLASSIFICATION OF SOIL WATER

Soil water has been classified from a physical and biological point of view as Physical classification of soil water, and biological classification of soil water.

Physical classification of soil water

i) Gravitational water ii) Capillary water and iii) Hygroscopic water

i. Gravitational water: Gravitational water occupies the larger soil pores (macro pores) and moves down readily under the force of gravity. Water in excess of the field capacity is termed gravitational water. Gravitational water is of no use to plants because it occupies the larger pores. It reduces aeration in the soil. Thus, its removal from soil is a requisite for optimum plant growth. Soil moisture tension at gravitational state is zero or less than 1/3 atmosphere.

Factors affecting gravitational water

i. Texture: Plays a great role in controlling the rate of movement of gravitational water. The flow of water is proportional to the size of particles. The bigger the particle, the more rapid is the flow or movement. Because of the larger size of pore, water percolates more easily and rapidly in sandy soils than in clay soils.
ii. Structure: It also affects gravitational water. In platy structure movement of gravitational water is slow and water stagnates in the soil. Granular and crumby structure helps to improve gravitational water movement. In clay soils having single grain structure, the gravitational water, percolates more slowly. If clay soils form aggregates (granular structure), the movement of gravitational water improves. Capillary water: Capillary water is held in the capillary pores (micro pores).

2. **Capillary water** is retained on the soil particles by surface forces. It is held so strongly that gravity cannot remove it from the soil particles. The molecules of capillary water are free and mobile and are present in a liquid state. Due to this reason, it evaporates easily at ordinary temperature though it is held firmly by the soil particle; plant roots are able to absorb it. Capillary water is, therefore, known as available water. The capillary water is held between 1/3 and 31 atmosphere pressure.

Factors affecting capillary water

The amount of capillary water that a soil is able to hold varies considerably.

The following factors are responsible for variation in the amount of capillary water.

   i. Surface tension: An increase in surface tension increases the amount of capillary water.
   ii. Soil texture: The finer the texture of a soil, greater is the amount of capillary water holds. This is mainly due to the greater surface area and a greater number of micro pores.
   iii. Soil structure: Platy structure contains more water than granular structure.
   iv. Organic matter: The presence of organic matter helps to increase the capillary capacity of a soil. Organic matter itself has a great capillary capacity.

Undecomposed organic matter is generally porous having a large surface area, which helps to hold more capillary water. The humus that is formed on decomposition has a great capacity for absorbing and holding water. Hence the presence of organic matter in soil increases the amount of capillary water in soil.

3. **Hygroscopic water:** The water that held tightly on the surface of soil colloidal particle is known as hygroscopic water. It is essentially non-liquid and moves primarily in the vapour form.

   Hygroscopic water held so tenaciously {31 to 10000 atmosphere) by soil particles that plants cannot absorb it. Some microorganism may utilize hygroscopic water. As hygroscopic water is held tenaciously by surface forces its removal from the soil requires a certain amount of energy. Unlike capillary water which evaporates easily at atmospheric temperature, hygroscopic water cannot be separated from the soil unless it is heated

Factors affecting hygroscopic water:

Hygroscopic water is held on the surface of colloidal particles by the dipole orientation of water molecules. The amount of hygroscopic water varies inversely with the size of soil particles. The smaller the particle, the greater is the amount of hygroscopic water it adsorbs.

Fine textured soils like clay contain more hygroscopic water than coarse - textured soils. The amount of clay and also its nature influences the amount of hygroscopic water. Clay minerals of the montmorillonite type with their large surface area adsorb more water than those of the kaolinite type, while illite minerals are intermediate.

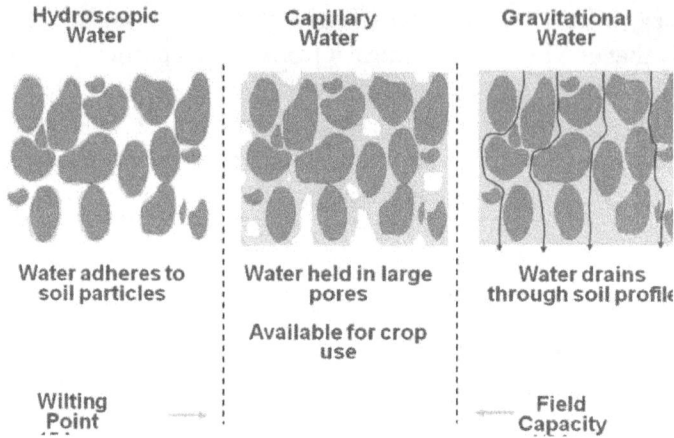

Hydroscopic Water	Capillary Water	Gravitational Water
Water adheres to soil particles	Water held in large pores	Water drains through soil profile
	Available for crop use	
Wilting Point		Field Capacity

Fig. 2

Biological Classification of Soil Water

There is a definite relationship between moisture retention and its utilization by plants. This classification based on the availability of water to the plant. Soil moisture can be divided into three parts.

    i. Available water: The water which lies between wilting coefficient and field capacity. It is obtained by subtracting wilting coefficient from moisture equivalent.

    ii. Unavailable water: This includes the whole of the hygroscopic water plus a part of the capillary water below the wilting point.

    iii. Super available or superfluous water: The water beyond the field capacity stage is said to be super available. It includes gravitational water plus a part of the capillary water removed from larger interstices. This water is unavailable for the use of plants. The presence of super-available water in a soil for any extended period is harmful to plant growth because of the lack of air.

5

STRESS DISTRIBUTION IN SOILS

INTRODUCTION

Soil mass is generally a three phase system that consists of solid particles, liquid and gas. The liquid and gas phases occupy the voids between the solid particles as shown in Figure 1a. For practical purposes, the liquid may be considered to be water (although in some cases the water may contain some dissolved salts or pollutants) and the gas as air. Soil behavior is controlled by the interaction of these three phases. Due to the three phase composition of soils, complex states of stresses and strains may exist in a soil mass. Proper quantification of these states of stress, and their corresponding strains, is a key factor in the design and construction of transportation facilities.

The first step in quantification of the stresses and strains in soils is to characterize the distribution of the three phases of the soil mass and determine their inter-relationships. The inter-relationships of the weights and volumes of the different phases are important since they not only help define the physical make-up of a soil but also help determine the in-situ geostatic stresses, i.e., the states of stress in the soil mass due only to the soil's self-weight. The volumes and weights of the different phases of matter in a soil mass shown in Fig. 1a can be represented by the block diagram shown in Fig. 1b. Such a diagram is also known as a phase diagram. A block of unit cross sectional area is considered. The symbols for the volumes and weights of the different phases are shown on the left and right sides of the block, respectively. The symbols for the volumes and weights of the three phases are defined as follows:

- ◉ V_a, W_a : volume, weight of air phase. For practical purposes, $W_a = 0$.
- ◉ V_w, W_a: volume, weight of water phase.
- ◉ V_v, W_a : volume, weight of total voids. For practical purposes, $W_v = W_w$ as $W_a = 0$.
- ◉ V_s, W_s : volume, weight of solid phase.
- ◉ V, W : volume, weight of the total soil mass .

Although $W_a = 0$ so that $W_a = W_a$, V_a is generally > 0 and must always be taken into account. Since the relationship between V_a and V_a usually changes with groundwater conditions as well as under imposed loads, it is convenient to designate all the volume not occupied by the solid phase as void space, Vv. Thus, $V_a = V_a + V_a$. Use of the terms illustrated in Fig. 1b, allows a number of basic phase relationships to be defined and/or derived as discussed next.

BASIC WEIGHT-VOLUME RELATIONSHIPS

Various volume change phenomena encountered in geotechnical engineering, e.g., compression, consolidation, collapse, compaction, expansion, etc. can be described by expressing the various volumes illustrated in Fig. 1b as a function of each other. Similarly, the in-situ stress in a soil mass is a function of depth and the weights of the different soil elements within that depth. This in-situ stress, also known as overburden stress, can be computed by expressing the various weights illustrated in Fig. 1b as a function of each other. This section describes the basic inter-relationships among the various quantities shown in Fig. 1b.

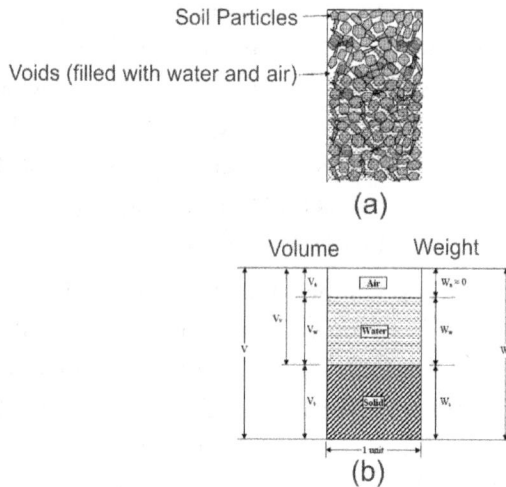

(a)

(b)

Fig. 1. A unit of soil mass and its idealization.

Volume Ratios

A parameter used to express of the volume of the voids in a given soil mass can be obtained from the ratio of the volume of voids, V_v, to the total volume, V. This ratio is referred to as porosity, n, and is expressed as a percentage as follows:

$$n = \frac{V_v}{V} \times 100 \qquad (1)$$

Obviously, the porosity can never be greater than 100%. As a soil mass is compressed, the volume of voids, V_v, and the total volume, V, decrease. Thus, the value of the porosity changes. Since both the numerator and denominator in Equation 1 change at the same time, it is difficult to quantify soil compression, e.g., settlement or consolidation, as a function of porosity. Therefore, in soil mechanics the volume of voids, V_v, is expressed in relation to a quantity, such as the volume of solids, V_s, that remains unchanging during consolidation or compression. This is done by the introduction of a quantity known as void ratio, e, which is expressed in decimal form as follows:

$$e = \frac{Vv}{Vs} \tag{2}$$

Unlike the porosity, the void ratio can have values greater than 1. That would mean that the soil has more void volume than solids volume, which would suggest that the soil is "loose" or "soft." Therefore, in general, the smaller the value of the void ratio, the denser the soil. As a practicality, for a given type of coarse-grained soil, such as sand, there is a minimum and maximum void ratio. These values can be used to evaluate the relative density, D_r (%), of that soil at any intermediate void ratio as follows:

$$D_r = \frac{(e_{max} - E)}{(e_{max} - e_{min})} \tag{2a}$$

At $e = e_{max}$ the soil is as loose as it can get and the relative density equals zero. At $e = e_{min}$ the soil is as dense as it can get and the relative density equals $\%100$. Relative density and void ratio are particularly useful index properties since they are general indicators of the relative strength and compressibility of the soil sample, i.e., high relative densities and low void ratios generally indicate strong or incompressible soils; low relative densities and high void ratios may indicate weak or compressible soils.

While the expressions for porosity and void ratio indicate the relative volume of voids, they do not indicate how much of the void space, V_v, is occupied by air or water. In the case of a saturated soil, all the voids (i.e., soil pore spaces) are filled with water, $V_v = V_w$. While this condition is true for many soils below the ground water table or below standing bodies of water such as rivers, lakes, or oceans, and for some fine-grained soils above the ground water table due to capillary action, the condition of most soils above the ground water table is better represented by consideration of all three phases where voids are occupied by both air and water. To express the amount of void space occupied by water as a percentage of the total volume of voids, the term degree of saturation, S, is used as follows:

$$S = \frac{V_w}{V_v} \times 100 \tag{3}$$

Obviously, the degree of saturation can never be greater than 100%. When S = 100%, all the void space is filled with water and the soil is considered to be saturated. When S = 0%, there is no water in the voids and the soil is considered to be dry.

Weight Ratios

While the expressions of the distribution of voids in terms of volumes are convenient for theoretical expressions, it is difficult to measure these volumes accurately on a routine basis. Therefore, in soil mechanics it is convenient to express the void space in gravimetric, i.e., weight, terms. Since, for practical purposes, the weight of air, W_a, is zero, a measure of the void space in a soil mass occupied by water can be obtained through an index property known as the gravimetric water or moisture content, w, expressed as a percentage as follows:

$$w = \frac{W_w}{W_s} \times 100 \tag{4}$$

The word "gravimetric" denotes the use of weight as the basis of the ratio to compute water content as opposed to volume, which is often used in hydrology and the environmental sciences to express water content. Since water content is understood to be a weight ratio in geotechnical engineering practice, the word "gravimetric" is generally omitted. Obviously, the water content can be greater than 100%. This occurs when the weight of the water in the soil mass is greater than the weight of the solids.

In such cases the void ratio of the soil is generally greater than 1 since there must be enough void volume available for the water so that its weight is greater than the weight of the solids. However, even if the water content is greater than 100%, the degree of saturation may not be 100% because the water content is a weight ratio while saturation is a volume ratio.

For a given amount of soil, the total weight of soil, W, is equal to $W_s + W_w$, since the weight of air, W_a, is practically zero. The water content, w, can be easily measured by oven-drying a given quantity of soil to a high enough temperature so that the amount of water evaporates and only the solids remain. By measuring the weight of a soil sample before and after it ahs been oven dried, both W and W_s, can be determined. The water content, w, can be determined as follows since $W_a = 0$:

$$w = \frac{W - W_s}{W_s} = \frac{W_w}{W_s} \times 100 \tag{4a}$$

Most soil moisture is released at a temperature between 220 and 230°F (105 and 110°C). Therefore, to compare reported water contents on an equal basis between various soils and projects, this range of temperature is considered to be a standard range.

THEORIES FOR DETERMINING VERTICAL STRESS IN SOIL MASS

When the structure is constructed on the ground, then the self-weight and live loads imposed over structure are transmitted to foundation. This foundation acts as connecting medium between the structure and the ground.

The load transmitted to the foundation is directly transferred to ground, through the foundation of the structure. This load transferred by the foundation applies the effort on the soil. This effort induces stresses in the soil. The distribution of these stresses is known as soil stress distribution.

The soil stress distribution affects the stability and safety of the structure. The factors that affect the soil stress distribution are as follow:

- The characteristics of the load,
- The geometric specifications of the foundation,
- The attributes of the soil and
- The method opted for the computation of stresses.

The soil stress distribution is caused by the self-weight of soil and surface loading on the structure.

Self-weight causes the stresses below the soil at any depth. The equation used to determine the stress distribution caused due to self-weight is as shown below:

$$\sigma_\varsigma = \gamma \times Z \tag{5}$$

Here, γ is unit weight, Z is soil depth below the ground surface and *ó z* is stress developed due to self-weight.

Consider Fig. 2 for the soil stress distribution caused by the self-weight.

Soil Stress Distribution due to surface loading from the structure: The surface stresses are developed due to the foundation placed on the surface. These stresses are calculated by two methods which are as follows:

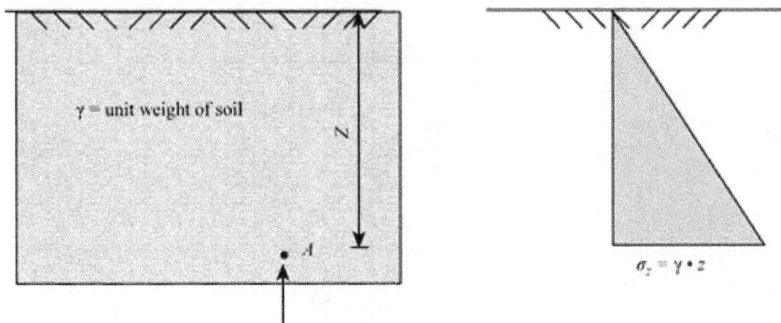

Fig. 2 Soil stress distribution caused by the self-weight

•**Boussinesq Formula:** The equation of stress distribution using Boussinesq formula is as given below:

$$\sigma_z = \frac{3P}{2\pi z^2[1 + (r/z)^2]^{5/2}} \tag{5}$$

Here, $s\,z$ is the stress developed in the soil, P is the concentrated load applied on the soil, z is the depth of the soil and r is the non-zero radius

•**Westergaard Formula:** The equation of stress distribution using Westergaard formula is as given below.

$$\sigma_\zeta = \frac{P}{\pi z^2[1 + 2\,(r/z)^2]^{3/2}} \tag{6}$$

Consider Figure 3 for the soil stress distribution due to surface load on the structure.

THEORIES FOR DETERMINING VERTICAL STRESS IN SOIL MASS

Boussinesq Theory for Vertical Stress due to Concentrated Load

Boussinesq in 1885 gave a solution for stress distribution in a homogeneous and isotropic subgrade subjected to a vertical concentrated load (point load) on the ground surface.

Boussinesq showed that for the point load acting at the ground surface, the polar stress σ_r at a point P as shown in Fig. 4 is given by –

$$\sigma_r = (3Q/2\pi)\,(\cos\theta/R^2) \tag{7}$$

where Q is the magnitude of the concentrated load, R is the radial distance of point P from the point of application of the load, and θ is the angle made by the radius R with the axis of the load. But –

$$\cos\theta = z/R \tag{8}$$

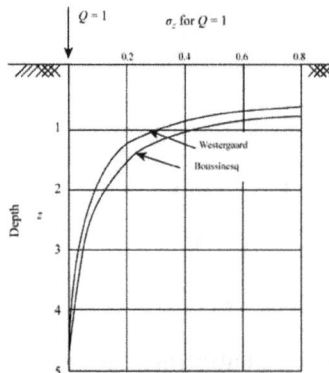

Fig. 3 The soil stress distribution due to surface load on the structure.

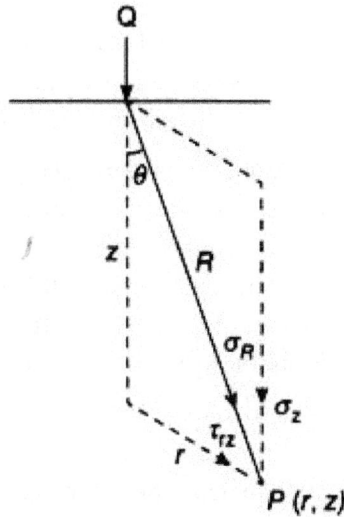

Fig. 4 Polar stress due to concentrated load

where z is the depth of point P below the ground level. Therefore –

$R = z/\cos\theta$ (9)

Substituting this value of R in Eq. (7), we have –

$\sigma_r = (3Q/2\pi)(\cos\theta/z/\cos\theta)^2 = (3Q/2\pi z^2)\cos^3\theta$ (10)

Vertical stress at point P is given by –

$\sigma_z = \sigma_r\cos^2\theta = (3Q/2\pi z^2)\cos^5\theta$ (11)

It may be noted that the Boussinesq influence factor for vertical stress is a function of the ratio r/z and is independent of soil properties. It is maximum when r = 0, that is, below the concentrated load, and is equal to 0.4775. Values of I_B are computed for different values of r/z. The vertical stress computed by the Boussinesq theory for a point load is infinite when z = 0, that is, at the ground surface, where the load is applied.

The following are the assumptions in the **Boussinesq theory**:

    a. The soil is homogeneous and isotropic.

    b. The soil mass is semi-infinite; that is, it extends infinitely in all directions below a level surface.

    c. The soil may not be elastic, but obeys Hooke's law; that is, the stress-strain relationship is linear.

    d. The soil is weightless and unstressed before application of the load.

    e. The load is applied at the ground surface.

Pressure Bulb

From the Boussinesq equation for vertical stress, it may be observed that the vertical stress decreases with the increase in depth from the ground surface. It also decreases with the increase in the distance from the axis of the load in the lateral direction at any depth. The zone of soil within which the vertical stress is significant and causes significant deformation of the soil, affecting the safety and stability of the structure, is known as pressure bulb. The soil outside the pressure bulb is assumed to have negligible stresses.

An isobar is a curve joining points of equal vertical stresses in the soil mass. Isobar is a spatial curved surface in the shape of an electric bulb or an onion. For a concentrated load acting at the ground surface, the isobar is symmetrical on either side of the axis of the load. A pressure bulb is an isobar of stress intensity 0.1Q, where Q is the magnitude of the concentrated load. The coordinates of the points on the pressure bulb can be obtained by substituting $\sigma_z = 0.1Q$ in the Boussinesq equation for vertical stress due to a point load.

The resulting coordinates of r and z are plotted to get the pressure bulb of intensity 0.1Q. It may be observed from the pressure bulb that r is zero directly below the point load and increases with the increase in depth below the ground surface.

The radial distance r attains a maximum value of 0.938 when z = 1.25. It then decreases with further increase in depth and becomes zero again at a depth of 2.185. The left-hand side of the pressure bulb can be drawn using the same value of r at different depths but with a negative sign. It is also possible to draw isobars of stress intensity 0.2Q, 0.5Q, and 0.75Q using a similar procedure.

Vertical Stress Distribution on a Horizontal Plane:

Thus, the vertical stress at a depth of 1 m can be computed in terms of Q by substituting different values of r in the above equation. The vertical stress distribution on a horizontal plane at depth 1 m is shown in Fig. 5. It may be seen that the vertical stress is maximum below the axis of the load equal to 0.4775Q and decreases rapidly to 0.27Q at r = 0.5 m and to 0.08Q at r = 1 m on the either side of the axis of the load.

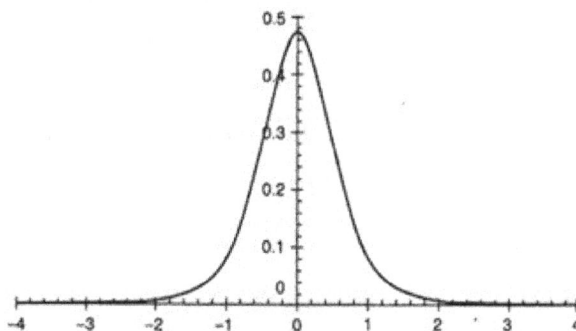

Fig. 5 Vertical stress distribution on a horizontal plane (based on Boussinesq for a point load at depth = 1 m)

Vertical Stress Distribution on a Vertical Plane

It is possible to determine the vertical stress distribution due to a point load on a vertical plane below the axis of the load by substituting r = 0. The vertical stress distribution on a vertical plane below the axis of the load is shown in Fig. 6. It starts from infinity, immediately below the load, and decreases very rapidly to an intensity of about 2Q within a depth of 0.5 m and to about 0.5Q at a depth of 1 m below load level. The distribution is asymptotic with the x-axis within this depth, whereas it is asymptotic with the y-axis over the remaining depth.

Vertical Stress Due to Line Load

A line load acts over negligible width for infinite length. The concept of line load is useful as it is extended to determine the vertical stress below strip footing.

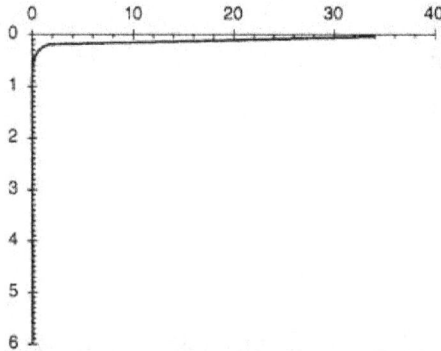

Fig. 6 Vertical Stress Distribution on a Vertical Plane

Consider a line load of infinite length having intensity q' per unit length. Consider a point P having coordinates (x, y, z) with respect to line load, at which the vertical stress is to be determined, as shown in Fig. 7. The point P is thus at a depth z below the load level and at a distance x normal to the length of the line load. The y-axis is along the length of the load.

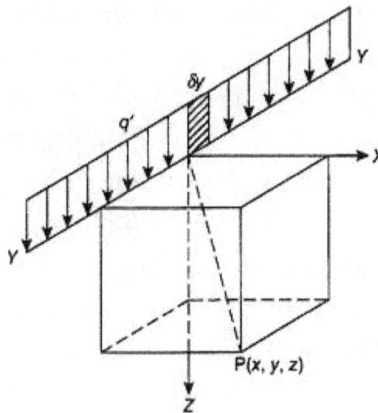

Fig. 7 Vertical Stress Due to Line Load

Consider an elemental length δy of the load. The total elemental load over this small length can be considered as concentrated load of Boussinesq's theory can be used to determine the elemental vertical stress at point P as –

$$\Delta\sigma_z = \frac{3\times(q'\times\delta y)}{2\pi} \times \frac{z^3}{(r^2 = z^2)^{5/2}} \tag{12}$$

WESTERGAARD'S THEORY FOR VERTICAL STRESS

The Boussinesq theory assumes that the soil mass is isotropic. Actual sedimentary soils are generally anisotropic. Thin layers of sand are usually embedded in a homogeneous clay deposit. Westergaard's theory assumes that thin sheets of rigid material are sandwiched in a homogeneous soil mass. These thin sheets are closely spaced and are of infinite rigidity; hence, they are incompressible. These thin sheets of sand permit only downward displacement of the soil mass without lateral deformation.

As per Westergaard's theory, vertical stress due to a point load is given by

$$\sigma_z = \frac{2q'}{\pi z}\left[\frac{1}{1+(x/z)^2}\right]^2 \tag{12}$$

or

$$\sigma_z = I_b\,(q'/z) \tag{13}$$

where σ_z is the vertical stress below the load Q at depth z at a radial distance r, I_w is Westergaard's influence factor, and μ is the Poisson's ratio of the soil. When $\mu = 0$

Thus, Westergaard's theory gives the vertical stress a 50% higher value than that given by the Boussinesq theory at any depth below the axis of the load.

THE 2:1 DISTRIBUTION METHOD:

For calculation of settlements, vertical stress is often obtained by assuming that the load is dispersed at a slope of 2 Vertical: 1 Horizontal from the base of the footing. Consider a rectangular footing ABCD of width b and length 1 subjected to a uniform pressure q. It is required to determine the vertical stress at depth z below the footing.

As per 2:1 distribution method, the resisting width AB at the foundation level will increase to EF at depth z, as shown in Fig. 8(a) and the resisting width BC at the foundation level will increase to FG at depth z, as shown in Fig. 8(b).

The resisting area ABCD at the foundation level will increase to EFGH at depth z, as shown in Fig. 8(c). Thus the load, which is acting over the area 1 × b foundation level, is dispersed at 2V:1H over depth z, so that the resisting area at depth z will be (1 + z) x (b + z) at depth z.

The vertical stress at depth z below the rectangular footing is therefore determined from –

$$\sigma_z = \frac{q\times(l\times b)}{(1 + z)(b + z)} \tag{13}$$

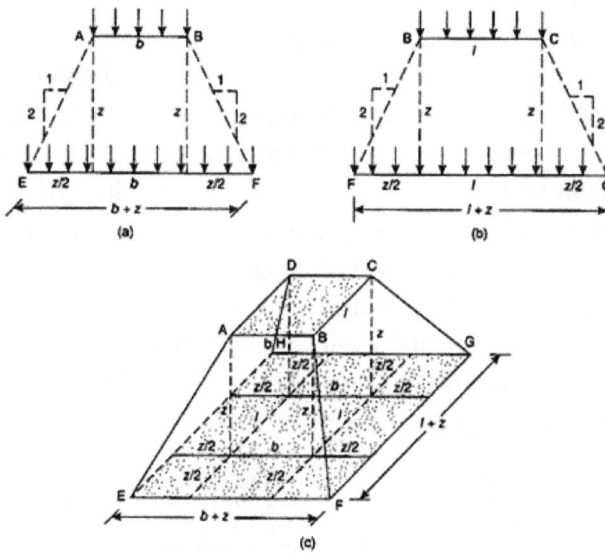

Fig. 8 (a) Load disperssion on side AB. (b) Load disperssion on side BC. and (c) Load disperssion below rectangular footing ABCD

The 2:1 distribution method enables easy and quick determination of vertical stress at any depth below a footing. The method is approximate but the error involved is not considerable and preliminary estimation of settlement of footings can be done using this method.

Comparison of Different Methods for Determination of Vertical Stress

Figure 9 shows the variation in vertical stress with depth as computed by the Boussinesq theory and the 2:1 approximate method. It may be observed from Fig. 9 that the vertical stress by the Boussinesq theory gives maximum and minimum values at the center and at the edge of the footing, respectively. The vertical stress computed by the 2:1 distribution method lies between these two extreme values at any depth. The average vertical stress is also shown in Fig. 9 with depth.

The vertical stress computed by the 2:1 distribution method is closer to the stress at the edge of the footing computed by the Boussinesq theory for depth > 0.5B. It approaches the average vertical stress at a depth of 0.2B. We know that the vertical stress at the edge of a square footing is about 50% of that at the center. Hence, the 2:1 distribution method underestimates vertical stress, especially for depths less than about 1.5B, where stresses are significant. Hence, the use of the 2:1 distribution method causes errors on the unsafe side.

Effect of Water on Physical States of Soils

For practical purposes, the two most dominant phases are the solid phase and the water phase. It is intuitive that as the water content increases, the contacts between the particles comprising the solid phase will be "lubricated." If the solid phase is comprised of coarse particles, e.g. coarse sand or gravels, then water will start flowing between the particles of the solid phase. If the solid phase is comprised of fine-grained particles, e.g., clay or silt, then water cannot flow as freely as in the coarse-grained solid phase because pore spaces are smaller and solids react with water. However, as the water content increases even the fine-grained solid phase will conduct water and under certain conditions the solid phase itself will start deforming like a viscous fluid, e.g., like a milk shake or a lava flow. The mechanical transformation of the fine-grained soils from a solid phase into a viscous phase is a very important concept in geotechnical engineering since it is directly related to the load carrying capacity of soils. It is obvious that the load carrying capacity of a solid is greater than that of water. Since water is contained in the void space, the effect of water on the physical states of fine-grained soils is important. Some of the basic index properties related to the effect of water are described next.

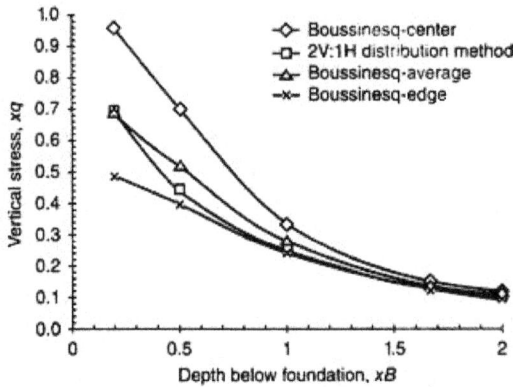

Fig. 9 Comparison of Different Methods for Determination of Vertical Stress

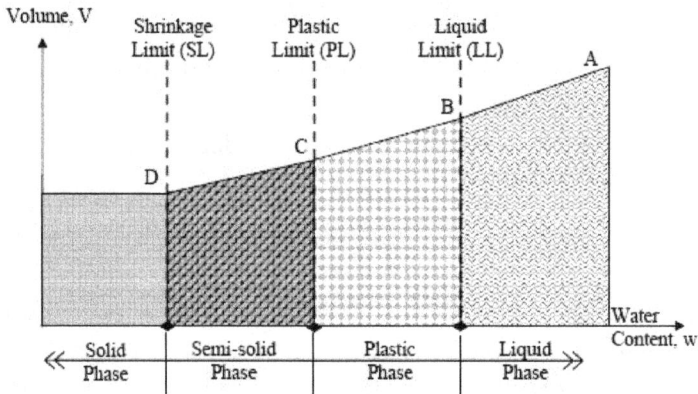

Fig. 10. Conceptual changes in soil phases as a function of water content.

The physical and mechanical behavior of fine-grained-soils is linked to four distinct states: solid, semi-solid, plastic and viscous liquid in order of increasing water content. Consider a soil initially in a viscous liquid state that is allowed to dry uniformly. This state is shown as Point A in Fig. 10, which shows a plot of total volume versus water content. As the soil dries, its water content reduces and consequently so does its total volume as the solid particles move closer to each other. As the water content reduces, the soil can no longer flow like a viscous liquid. Let us identify this state by Point B in Fig. 10. The water content at Point B is known as the "Liquid Limit" in geotechnical engineering and is denoted by LL.

As the water content continues to reduce due to drying, there is a range of water content at which the soil can be molded into any desired shape without rupture. In this range of water content, the soil is considered to be "plastic."

If the soil is allowed to dry beyond the plastic state, the soil cannot be molded into any shape without showing cracks, i.e., signs of rupture. The soil is then in a semi-solid state. The water content at which cracks start appearing when the soil is molded is known as the "Plastic Limit."

OVERBURDEN PRESSURE

Soils existing at a distance below ground are affected by the weight of the soil above that depth. The influence of this weight, known generally as overburden, causes a state of stress to exist, which is unique at that depth, for that soil. This state of stress is commonly referred to as the overburden or in-situ or geostatic state of stress. When a soil sample is removed from the ground, as during the field exploration phase of a project, that in-situ state of stress is relieved as all confinement of the sample has been removed. In laboratory testing, it is important to reestablish the in-situ stress conditions and to study changes in soil properties when additional stresses representing the expected design loading are applied.

The stresses to be used during laboratory testing of soil samples are estimated from either the total or effective overburden pressure. The engineer's first task is determining the total and effective overburden pressure variation with depth. This relatively simple task involves estimating the average total unit weight for each soil layer in the soil profile, and determining the depth of the water table. Unit weight may be reasonably well estimated from tests on undisturbed samples or from standard penetration N-values and visual soil identification.

The water table depth, which is typically recorded on boring logs, can be used to compute the hydrostatic pore water pressure at any depth. The total overburden pressure, p_t, is found by multiplying the total unit weight of each soil layer by the corresponding layer thickness and continuously summing the results with depth. The effective overburden pressure, p_o, at any depth is determined by accumulating the weights of all layers above that depth with consideration of the water level conditions at the site as follows:

Soils above the water table

⦿ Multiply the total unit weight by the thickness of each respective soil layer above the desired depth, i.e., $p_o = p_t$.

Soils below the ground water table

- Compute pore water pressure u as z_w γ_w where z_w is the depth below ground watertable and γ_w is the unit weight of water

- To obtain effective overburden pressure, p_o, subtract pore water pressure, u, from p_t

- For soils below the ground water table, p_t is generally assumed to be equal to p_{sat}.

EFFECTIVE STRESS IN SOIL: DEVELOPMENT, IMPORTANCE AND PRINCIPLES

Effective Stress Principle

The effective stress principle can be explained by studying an example of soil mass that is in a fully saturated condition as shown in Fig. 11 below.

Consider a prism of soil in the soil mass with an area of cross-section 'A'. If P is the weight of the soil over the soil prism element, 'P' is given by the formula:

P = Unit Weight x (Volume of the Prism)

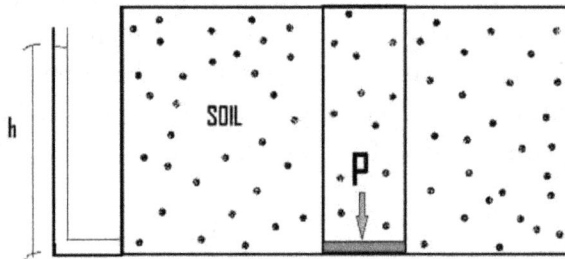

Fig.11. Saturated Soil Mass

i.e. P = Unit weight x Height of prism(h) x Area of Cross-section of Prism (A)

This gives the relation,

$P = \gamma_{sat} hA$

Effective Stress

Karl Terzaghi was the first to recognize the importance of effective stress. It is the stress transmitted through grain to grain at the point of contact through soil mass. It is also known as inter-granular stress. It is denoted by σ'. When soil mass is loaded. The load is transferred to the soil gains through their point of contact. If at the point of contact, the applied load is greater than the resistance of the grains, then there will be compression in the soil mass.

This compression is partly due to the elastic compression of the grains at the points of contact and partly due to relative sliding between particles. This load per unit area of soil mass responsible for deformation of the soil mass is termed as effective stress.

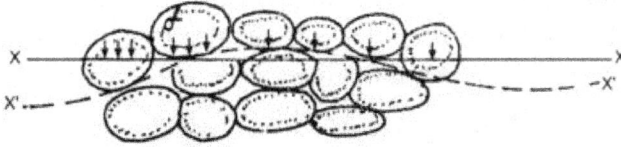

Fig. 12. Load transmitted

Neutral Stress

It is the stress or pressure transmitted through the pore fluid. It is also termed as pore pressure and is denoted by u. In saturated soil, pores of the soil mass are filled with water. When the saturated soil mass is loaded, the load is not transmitted through the grains. The load is transferred to the pore water. As water is incompressible, a pressure is developed in the pore water.

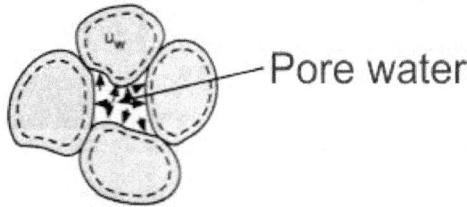

Fig. 13. Neutral Stress

This pressure is called pore pressure or pore water pressure. Pore pressure does not have any measurable influence on the mechanical property of the soil like void ratio, shear strength etc. This pressure or stress is called neutral stress.

Total Stress

Total stress is equal to the sum of the effective stress and the neutral stress. It is denoted by σ.

$\sigma = \sigma + u$

Effective stress cannot be measured in the field by any instrument. It can only be calculated after measuring total stress and pore pressure. Thus effective stress is not a physical parameter, but is only very useful mathematical concept for determination of engineering behaviour of soil.

Importance of Effective Stress in Engineering Problems:

The effective stress plays an important role in:

- ◉ (i) Settlement of soil
- ◉ (ii) Shear strength of soil

Settlement of Soil:

The phenomenon of gradual reduction in volume of soil due to expulsion of water from soil pores is called consolidation or compression or settlement of soil. Figure 14 shows a compression curve of clay. It is a curve between effective stress σ and void ratio e. It is clear from the graph that when ó increases e decreases i.e., due to increase in the effective stress the compression of soil will increase.

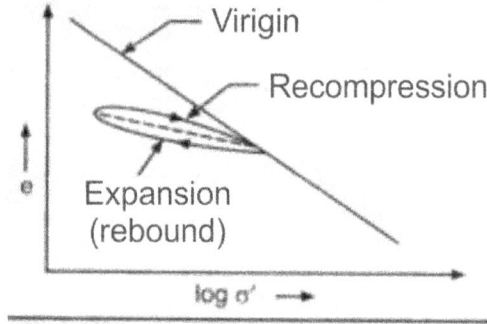

Fig. 14 Compression curve of clay

The final consolidation settlement may be calculated by using the formula

$$S = mV \, H \, \Delta\sigma$$

where mV is the coefficient of volume compressibility

H is the thickness of compressible layer

σ is the average increase in effective pressure.

From the above equation it is clear that settlement of soil is directly proportional to the effective pressure. So the settlement of soil depends upon the effective stress or effective pressure. As the effective stress increases, the settlement of the soil also increases.

Shear Strength of Soil

Many geotechnical engineering problems require an assessment of shear strength including:

a. **Structural foundations:** Load from a structure is transferred to ground through foundation. This produces shear stress and compressive stress. If shear stress produced is more than the shear strength of soil, shear failure occurs which cause the structure to collapse.

b. **Earth slopes:** On a sloping ground, gravity produces shear stresses in the soil. If these stresses exceed the shear strength, a landside occurs.

c. **Highway pavements:** Wheel loads, from vehicles are transferred through pavement to the ground. These loads produce shear stress which causes shear failure.

6

SHEAR STRENGTH OF SOILS

INTRODUCTION

The shear strength of soil determines its resistance to deformation by tangential (or shear) stress. Soil that has greater shear strength will have more cohesion between particles and more friction or interlocking to prevent particles sliding over each other. Soil shear strength is used to calculate bearing capacity and design retaining walls, slopes and embankments.

The strength of soil is typically defined as the resistance to shear stress in terms of the effective internal friction angle and effective cohesion (c'). This technical note is intended to provide a basis for estimating a soil's unit weight in the absence of specific testing results.

Soil shear strength depends on many elements, one fundamental consideration being the type of soil. The particles of most soils are essentially incompressible, and soil masses have no tensile strength. Soils fail when one block of soil moves relative to another block and the soil particles at the failure plane move over each other.

This is what is known as shear. When particles move across each other, the resisting (or shear) force is friction. The shear resistance, or shear strength of soil, is related to the physical characteristics of the soil, including particle size, shape, distribution and orientation, but also the stresses acting on the soil at that location.

IMPORTANCE OF SHEAR STRENGTH OF SOIL

The knowledge of shear strength is very important some of the uses are provided below:

- ◉ In the design of foundations the evaluation of bearing capacity is dependent on the shear strength.

- ◉ For the design of embankments for dams, roads, pavements, excavations, levees etc. The analysis of the stability of the slope is done using shear strength.

- In the design of earth retaining structures like retaining walls, sheetpile coffer dams, bulks heads, and other underground structures etc.

- The shear strength of a soil mass is essentially made up of:

- Due to the interlocking of the grains the structural resistance of the movement of the soil is very essential.

- An other important component is the frictional resistance between the individual soil grains at their contact point on sliding.

- The resistance due to the forces which hold the particles together or the cohesion.

FACTORS AFFECTING SOIL SHEAR STRENGTH

The shear strength of a soil is achieved by interaction between the solid, liquid, and gas particles in its make-up. So the shear strength of a soil depends on the composition of the soil's particles, the amount of water in the soil, and how well compacted the soil is. The contributing factors include, but are not limited to:

- Mineralogy of the soil particles (e.g. silica, quartz, feldspar, etc.).

- The range of sizes of the soil particles, also known as the particle size distribution.

- The angularity of the soil particles (most relevant to coarse sands and gravels)

- The moisture content of the soil – whether the voids between the soil particles are completely filled with water (fully saturated) or mostly air – and the capillary forces created by the interaction of the solid particles, water and air.

- Degree of compaction of the soil.

MEASURING SOIL SHEAR STRENGTH

The shear strength of a soil is measured directly in a laboratory or estimated from correlations with testing undertaken on site. In a laboratory, shear strength is measured by shear box or triaxial testing in accordance with BS 1377-7:1990 or BS 1377-8.:1990 respectively.

The following are the various types of tests available for the determination of shear strength:

1. Direct Shear Test
2. Triaxial Compression Test.
3. Unconfined Compression (UCC) Test.
4. Vane Shear Test.
5. Bore Hole Shear Test.

Direct Shear Test:

Direct shear test is a simple and commonly used test performed in a shear box to determine the shear parameters of soils. The direct shear test is also known as shear box test.

The principle of the test is to cause shear failure of a soil specimen, placed in a shear box along a predetermined horizontal plane, under a given normal stress, and to determine the shear stress at failure. The test is repeated on identical soil specimens under different normal stresses and the shear stress at failure under each normal stress is determined. A graph is plotted between the normal stress and the shear stress and the y-intercept and the slope of the failure envelope so obtained are taken as the shear parameters c and ϕ, respectively.

Triaxial Compression Test:

Triaxial compression test is used for the determination of shear parameters of all types of soil under any drainage condition.

In this test, a cylindrical soil specimen is subjected to a confining pressure in all directions in a triaxial cell. An additional axial load, known as deviator load, is then applied in vertical direction until the specimen fails (refer to Fig. 1). Here σ_3 is the confining pressure, also called cell pressure, or all-round pressure and σ_d is the deviator stress, also called additional axial stress or added axial stress.

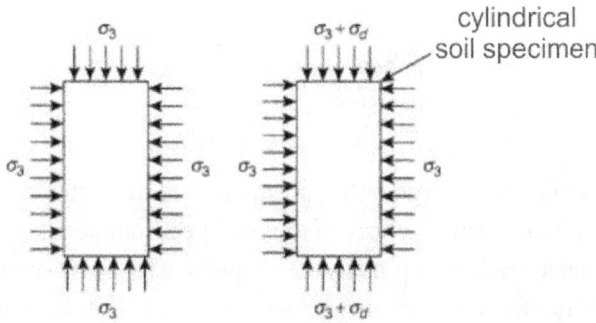

Fig. 1 Triaxial compression test: (a) consolidated stage and (b) shear stage

The test is repeated on three or more identical soil specimens and the principal stresses, so obtained, are used to draw the Mohr's circles. The failure envelope is drawn as a common tangent for the Mohr's circles to determine the shear parameters of the soil.

Unconfined Compression Test

The unconfined compression (UCC) test is a special case of a triaxial compression test in which the confining (cell) pressure is zero. The test can be conducted only on saturated cohesive soils, which can stand unsupported without confining pressure. A cylindrical soil specimen is subjected to axial vertical stress (major principal stress), as shown in Fig. 2, until the specimen fails due to shearing along a critical plane of failure.

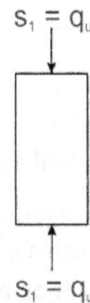

Fig. 2. Soil specimen unconfined compression (UCC) test

The merits of the UCC test are that the test is simple, convenient, and quick. It is ideally suited for measuring undrained shear strength of saturated clays.

In its simplest form, the apparatus consists of a load frame fitted with a proving ring to measure the vertical stress, s1 applied to the soil specimen. Figure 3 shows the test setup for the UCC test. The deformation of the soil specimen during the application of vertical stress is measured by a separate dial gauge. The load versus deformation readings during the test are taken and the stress-strain curve is plotted.

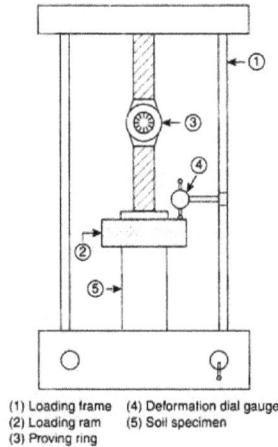

(1) Loading frame   (4) Deformation dial gauge
(2) Loading ram   (5) Soil specimen
(3) Proving ring

Fig. 3 UCC test setup

The soil specimen may undergo either plastic failure or brittle failure in the UCC test, as shown in Fig. 4. When a brittle failure of the soil specimen occurs, the proving ring dial gauge indicates a peak load, which decreases rapidly with further increase of strain. In brittle failure, a distinct failure plane in the form of a crack will occur across the specimen as shown in Fig. 4.

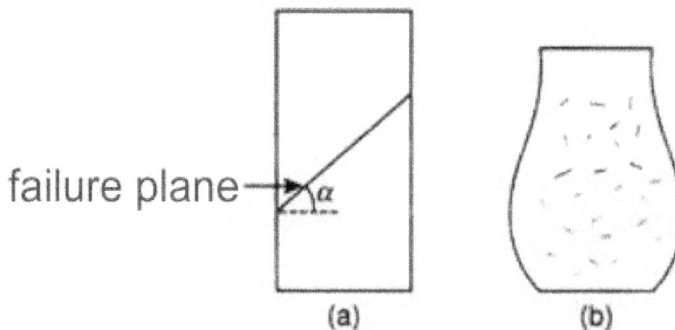

failure plane α

(a)   (b)

Fig. 4. Soil specimen in UCC test: (a) Brittle failure and (b) Plactic failure

In plastic failure, no definite peak load is indicated and the soil specimen bulges laterally. In such a case, the load corresponding to 20% axial strain is taken as the failure load. There will be no distinct failure plan and a number of small and fine cracks occur, distributed throughout the specimen.

Figure 5 shows the stress conditions in the soil specimen at the time of failure. The test is essentially an undrained test, if it is assumed that no moisture content is lost from the soil specimen during the test. Thus, for saturated cohesive soils in UCC test.

$\sigma_3 = 0$ and $\Phi u = 0$

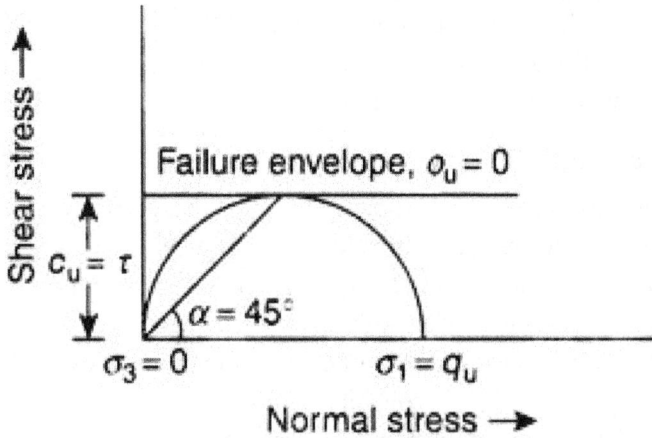

Fig. 5 Mohe's circle for saturated chhesive soil in UCC test

Thus, the Mohr's circle of stress starts from the origin and the diameter of the Mohr's circle will be $\sigma_1 = \Phi_u$, where qu is the unconfined compressive strength of the soil.

Vane Shear Test

The laboratory vane shear test is used for the measurement of undrained shear strength of cohesive soils of low shear strength less than about 0.5 kgf/cm2. This test gives the undrained strength of the soil, and the undisturbed and remolded strengths obtained are also used for evaluating the sensitivity of the soil.

Bore Hole Shear Test

It is a field test to determine the in situ undrained strength of soft clays, in which the bore hole shear device is inserted into the bore hole and expanded into the soil using compressed air. The soil is sheared and the corresponding load Pv is determined. The test device operates by applying the direct shear test principle in situ. The results from the bore hole shear test approximates a CU test.

FRICTION

A key element of soil shear strength is friction. For objects in contact, friction force along a plane varies with the pressure acting perpendicular to the plane (known as normal stress). As normal stress increases, so too does frictional resistance, or shear stress. This underlines that soil shear strength is not a single figure but is dependent upon the stresses that are acting upon a soil. For granular soils, the relationship between shear stress and normal stress is a straight line, defined by an angle (ø) known as the friction angle.

Friction angle

When considering soil shear strength, it is important to know the friction angle, as well as the stresses which will be acting on the soil. When comparing different types of granular soil, the friction angle is the single property that defines strength. The friction angle is critical when sourcing materials for structural fill in reinforced soil structures and important for applications such as working platforms and road foundations. It must also be remembered that any change in friction angle may necessitate a re-design.

Clay soil shear strength

Clay soils are also composed of particles, although the particles are extremely small. There are electrostatic charges (attractive forces) acting between these fine particles, and surface tension from pore water holds particles together even without the application of external confining forces, hence clay soils have some shear strength even when normal stress is zero. This additional strength is known as apparent cohesion. It is not, however, a fundamental soil property. When considering the shear strength of granular soils, even those with some clay content, cohesion can mostly be ignored as it is the friction angle which is key.

The influence of groundwater on the shear strength of soil

The presence of groundwater also has a marked influence on soil shear strength. Soils can be saturated, where all void spaces between particles are filled with water, or partially saturated when a percentage of air bubbles are present within the void spaces. The pore pressures generated affect particle to particle stress and therefore friction between soil particles. When a load is applied to saturated soil, the pressure of the pore water (water in the spaces) immediately increases because water is incompressible.

Granular soils have relatively large interconnecting voids between particles. When – as is usually the case in construction – loads are applied slowly onto a granular soil, water can be assumed to drain away freely, dissipating any increased pore pressure and allowing the applied load to be transferred to the soil skeleton. Consequently, pore water effects can be ignored.

Clay soil reacts differently as the voids are microscopically small and poorly interconnected. Water can therefore only move at much lower rates and drainage is very slow, so when a load is applied to clay soil, the pore water pressure is unable to dissipate. As water is incompressible, the pore water pressure carries the load and the particle-to-particle stresses in the clay do not increase. The short-term saturated shear stress of clay soil is therefore a constant value referred to as the undrained stress, denoted by cu or su.

When considering the bearing capacity of clay soils, the undrained shear strength is critical. This is extremely common in the UK where soils with high clay content are found on many construction sites. Over time, water will drain from clay and the pore pressure will slowly reduce, therefore the strength of the clay soils will increase – however this is a very long-term effect.

The impact of soil state on the shear strength of soil

A further factor affecting granular soil shear strength is the degree of particle compaction or soil state. When a load is applied to loose, uncompacted soil, the particles will move closer together as the soil contracts. After contraction of the soil, shearing takes place as particles begin to move over each other. The soil shear strength increases as the particles compact, ultimately continuing at a constant level and constant density or volume.

Where granular soils are already densely compacted, little or no contraction takes place and particles are interlocked. As the load increases, before particles can shear over one another, they have to move apart along the shear plane, unlocking the interlock. This is known as dilation. The shear force required to overcome dilation is the 'peak strength' (øpeak). After dilation, particles can move over each other more easily, requiring a lower shear force than at peak, which is referred to as strength at constant volume øcv.

In a situation where the soil is not expected to shear – for example, the structural fill in a reinforced soil wall – the peak strength should be used in the design. If addressing a high deformation condition, the preferred option would more likely be the shear strength of the soil at constant volume. The important consideration is to ensure that the materials report provides the appropriate strength for the design purpose.

TRACTION

Introduction

Of the three principal ways of transmitting tractor-engine power into useful work - power takeoff, hydraulic, and drawbar - the least efficient and most used method is the drawbar. Traction is the term applied to the driving force developed by a wheel, track, or other traction device.

Traction in soils

Soils, like metals, can behave both elastically and plastically. Elastic deformation refers to the ability of the deformed material to return to its original dimensions. Plastic deformation refers to a condition of permanent deformation. For a soil in the elastic condition, a given applied force causes a known deformation. On removal of the force, recovery takes place.

If, however, the force is continually increased, a loading condition will occur that will cause the soil to deform permanently, i.e., it will behave plastically. The onset of this plastic condition is generally considered to be induced by shear failure, i.e., the sliding of one particle over another. In this case, the ability of a particular soil to support a given load before a permanent change to the soil structure occurs is called the shear strength of the soil.

Shear strength

Granular materials such as soils exhibit cohesive and frictional properties. Cohesion is the bonding together of soil elements irrespective of the type of applied load. Pure clay fits this category; dry sand, on the other hand, exhibits a frictional resistance to shear loads in that the resistance to shear increases with applied load. Most soils exhibit a combination of cohesive and frictional properties. The water content in clay soil has a strong influence on shear strength: the higher the water content, the lower the shear strength. An applied load will result in a wedge-shaped cone below the tire, which will cause deformation and displacement of adjacent soil particles. If the resultant shear stresses are greater then the soil can sustain, sinkage will occur, increasing the surface area. This will continue until the soil is able to support the tire and load.

Two forms of shear strength may be considered:

  i. **Bulk shear strength:** the resistance offered to movement by a relatively large volume of soil aggregates.
  ii. Clod shear strength: the resistance offered by the individual clod or aggregate.

The main factors that influence shear strength are:

  i. Moisture content
  ii. Packing density and particle size
  iii. Organic matter content.

Effect of agricultural traffic over soil

Mechanization of the agricultural processes and the use of heavy productive units, with high working width, resulted, over time, in soil compaction and reduction of harvest. The use of the organic fertilizers and deep tillage finally results in the diminishing of the content of organic matter and in an increased sensibility to compaction.

Soil compaction processes. Soil compaction may be defined as "the compaction of soil mass in a smaller volume". The increase of the bulk density is accompanied by structural changes, changes in the thermal and hydraulic conductivity and in the gas transfer characteristics, all of them affecting the chemical and biological equilibrium of soil.

The artificial (anthropic) soil compaction is the result of exaggerated traffic imposed by agricultural operations, transport operations etc. The intensity of anthropic compaction depends on different factors. Some of them belong to soil, namely to its susceptibility to compaction: uneven grain size distribution, unstable structure, reduced humus content etc. Other factors are influenced by the characteristics of the agricultural equipment; this compaction is favored by the heavy machineries exerting high pressure over soil, by the increased number of passings, by the increased tire air pressure, by the agricultural traffic performed over wet soil etc.

The effects of soil compaction. Soil compaction leads to the reduction of harvest by 50%, compared to the non-compacted soil, while fuel consumption is expected to be increased by 35%. Soil compaction is one of the main causes of surface flow and erosion. In the meantime, compacted soils require higher costs of the irrigation arrangements and exploitation, due to the poor infiltration of water and to the intensified evapotranspiration.

Soil compaction also causes the reduction of the water holding capacity and of the permeability, reduces soil aeration, significantly increases the penetration resistance and plowing resistance, inhibits the development of the plant roots and the quality of the ploughland deteriorates. It was established that not only the active parts of the tillage equipment deteriorate the soil structure, but also the tractors' wheels and the tracking wheels of the agricultural machinery; the use of heavier equipments leads to contact pressures of 0.2-1.8 MPa, while the specific resistance of the soil's structure elements is lower than 0.1 MPa (usually 0.02-0.006 MPa); as a result, soil compaction occurs up to 30-50 cm under the tracking wheel and on an area with a width four times the wheel width.

Plant and response to compaction. In compacted soils or compacted layers the penetration depth of the roots and their density are restricted, the effect being a slow development of the root system. As a result, the plant's access to water and nutrients is limited, while the capacity of the root system to counteract the noxious effect of pathogenic agents is diminished.

This is why plant species with deep roots are less sensible to compaction. Compaction induces changes in the soil's water and air regime, also affecting the activity of microorganisms. Soil compaction also favors the ammonia nitrogen in the detriment of nitric nitrogen, with unfavorable effects over the harvest.

Effect of soil tillage over compaction. The agricultural operations related to soil are: ploughing, land preparation, seeding and some of the maintenance operations. In most of the cases, the aim of these operations is to loosen the compacted soil layers. Where soil compaction is a problem, tillage has an ameliorative effect. Soils are usually subjected to two types of traffic: one that produces compaction (wheel traffic) and another one that produces loosening (tillage traffic).

Soil reaction to compaction depends on traffic characteristics, soil properties and humidity when the traffic takes place; soil compaction is usually expressed by the means of bulk density, porosity or penetration resistance. Because of wheeled traffic the bulk density of soil increases; the magnitude of the density change depends on the soil texture, its humidity, the wheel-soil contact pressure and the number of passages. The maximum compaction effect is reached when wheel slip reaches 15-25%, due to the fragmentation of the structure elements under the effect of shear stress.

Aggressiveness of the tillage active parts towards soil

Unlike the wheels (tracking wheels, driving wheels etc.), which compact the soil, the active parts of the tillage equipment (rotary cultivator tines, plow shares, cultivator tines etc.)

loosen the soil; the rollers, the combined seedbed preparation devices and the combined cultivator are exceptions.

In the case of the active parts used for seedbed preparation, their destructive action over the soil structure elements is of an utmost importance. The active parts destroy, to a greater or a lesser extent, the structure elements through deformation, crumbling, cutting, breaking-up. The destruction of the soil's structure is a general phenomenon, occurring at any tillage operation, but it gets large proportions in soils with a rough texture or average-rough texture (sandy soils, sand-loamy soils, clay-loamy soils) for which the mechanical stability of the structure elements is lower, due to the lower clay content.

The recently tilled soils, the humid soils or the dry clay soils on which loosening is obtained by the means of rotating active parts (disc harrows, rotary cultivators) are also vulnerable.

It should be emphasized that not only the active parts of the tillage equipments destroy the soil's structure, but also the wheels of the tractors or the tracking wheels of the agricultural machinery. In this case, the wheel-soil contact pressure produces compaction as a result of the deformation and breaking-up of the structure elements. The division of the structure elements into fragments results in the increase of the bulk density because of a more stuffed settlement.

Consequently, the soil's capacity to drain and store water is diminished, the thermal regime worsens, the accessibility of plants to nutrients is diminished and the activity of the anaerobic microorganisms is reduced; on sloping lands, the erosion due to water is intensified, and the plants have difficulties in developing the root system, loose their stability and harvest is diminished.

Effect of the moldboard plough over soil. In the working process, the plough's share cuts the furrow horizontally and begins its detachment, overturning and lateral displacement, the process being finalized by the moldboard. These actions cause complex deformations of the soil slice, resulting in it's the fragmentation and crumbling. It should be emphasized that, under the action of the share's cutting edge, a number of soil's structural elements are also cut. In the meantime, the twisting of the furrow, both vertically and horizontally, as well as the friction between soil and the parts of the plough body results in the destruction of a number of the structural elements through deformation, crumbling, breaking, fragmentation. The process of destruction of the structure elements is strengthened by the working speed and by the use of worn shares or aggressive moldboards (cylindrical or helical moldboards).

Effect of the rotary cultivators over soil. In the working process, due to the advance movement of the equipment and the rotation movement of the rotor, the active parts penetrate into the soil and cut slices with a particular shape. Under the action of centrifugal force, the soil slices are thrown over the inner surface of the housing and louver. As a result, a supplementary crumbling is achieved, the soil being left behind the rotary harrow in a loosened and fine layer.

The cutting edges of the tines of the rotary harrow cut some of the soil's structural elements, which are also destroyed through deformation, breaking, fragmentation, crumbling, due to the peripheral speed of the tines, to the friction between tines and soil and to the impact between the slices and the housing of the rotary harrow.

Pulverization of the structural elements may also occur, when some of them are fragmented to the maximum extent, resulting in particles of clay, silt and sand.

When using rotary harrows for the seedbed preparation care should be taken so that the width of the soil slices should not be less than 25 mm. In the working process, the number of destroyed structural elements increases when the slices get thinner. Therefore, the peripheral speed of the tines should no exceed 6 m/s, while the speed of the advance movement should not exceed 1 m/s.

Effect of the disc harrows over soil. The active parts of these agricultural equipments are spherical discs. In the tillage process, the disc is displaced forward, following the movement of the equipment; in the same time, the disc is rotating, due to the contact with the soil. Over the effect of weight, the disc penetrates into the soil and cuts a soil layer, which is raised over the interior concave surface of the disc, is crumbled, displaced laterally and partially overturned. The aggressiveness of the discs depends on their shape, as well as on the cutting angle (disk angle), which can be adjusted between 15 and 30 degrees; crumbling increases when the cutting angle increases.

During the working process, the cutting edge of the disc cuts some of the soil's structural elements; the number of cut structural elements is higher when the disk's cutting edge is indented. In the meantime, the bending of the soil layer, vertically and horizontally, its twisting, as well as the friction between soil and disk, cause the destruction of some structural elements, through deformation, breaking, fragmentation. The proportion of damaged structural elements increases with speed.

Effect of the cultivators over soil. Depending on their destination, cultivators are equipped with different types of active parts, the more important being the ones used for weed cutting and for soil loosening.

The weed cutting active parts are aimed to cut the weeds and loosen the soil. During the working process of the arrow type tines, soil is cut horizontally, at a certain depth; in the meantime, the weeds are cut and soil is loosened and crumbled.

The straight, chisel, diamond pointed and narrow arrow type loosening active parts are mounted on rigid or elastic holders. When elastic holders are used, the active part vibrates, exerting an energetic action over soil. As a result of the displacement of these active parts, the superficial soil layer is crumbled and loosened.

The weed cutting type active parts cut some of the soil's structural elements. All the types of active parts (for both weed cutting and soil loosening) destroy the structural elements due to the advance movement, friction, breaking of soil layers, vibration; the structural elements are destroyed through deformation, crumbling, breaking, fragmentation. In order to diminish these adverse effects, speed should be limited to 10-12 km/h.

STRENGTH PARAMETERS OF SOIL

Determination of Strength parameters

The shear strength parameters cohesion (c) and friction angle (F) can be determined by different laboratory tests for different types of soils.

Direct shear test

The soil sample is tested in a confined metal box of square cross-section. The box has two halves horizontally and a small clearance is maintained between them. Figure 6 shows a direct shear test set-up. Upper half of the box is fixed and lower half of the box is pushed or pulled horizontally with respect to the fixed half. Thus, a shear is applied in the soil sample. A constant normal force (vertical) is applied on the sample throughout the test. Then horizontal force or shearing is applied till the failure.

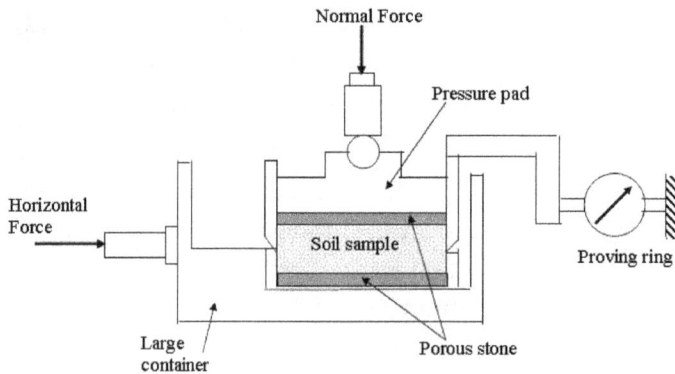

Fig. 6. Direct shear test

The shearing is normally applied at a constant rate of strain. The amount of shear load is measured with the help of proving ring. The vertical as well as horizontal deformation is measured with the help of dial gauges. The test procedure is repeated for different normal stresses (four to five normal stresses). The shear stress at failure is plotted against different normal stresses (as shown in Fig. 7). The shear strength parameters are determined from the bestfit straight line passing through the test points (as shown in Fig. 8). The test is suitable for sandy soils. If the sample is partially or fully saturated, porous stones are placed below and top of the sample to allow free drainage. Figure 8 shows typical shear stress-shear displacement and change in height of sample-shear displacement plot of soils obtained from direct shear tests.

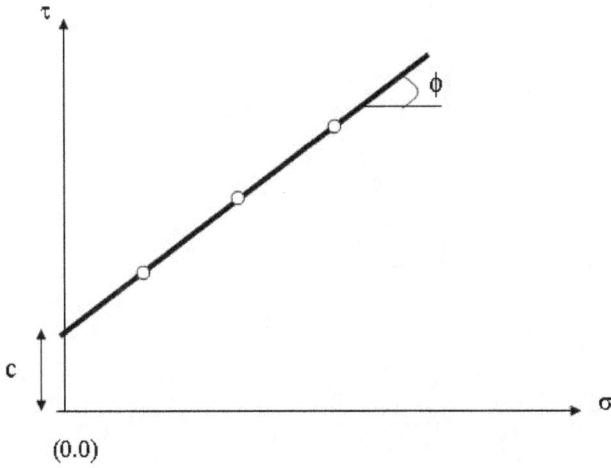

Fig. 7. Shear stress-normal stress plot

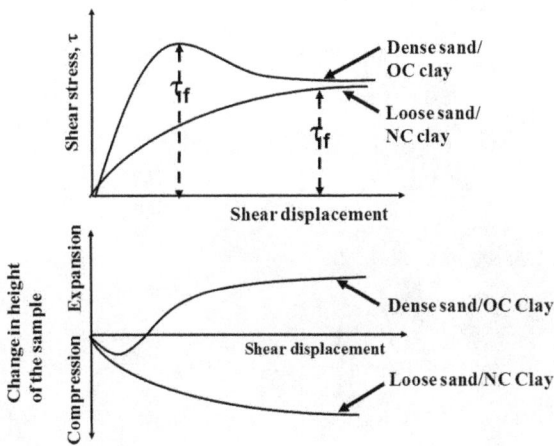

Fig. 8. Shear stress-shear displacement and change in height of sample-shear displacement plot of soils obtained from direct shear tests

7

STABILITY OF SLOPES

Slopes are typically categorized in two types: natural and artificially-made slopes. Natural slopes are formed due to physical processes that include plate tectonics and weathering/ erosion of rock masses that result in material deposition. Artificially-made slopes are established to facilitate infrastructure projects, ex.: embankments, earth dams, road cuttings etc.

The stability of a slope is of critical importance in Geotechnical Engineering applications. A slope movement (also referred as a landslide) can lead to severe issues including infrastructure damage or/and casualties. Slope stability depends on the capability of the soil mass to withstand its gravitational forces, the additional loads acting on the slope, as well as potential dynamic loads (such as that of an earthquake).

A common misconception is that landslides occur in steep and remote slopes and do not actually impact human infrastructure. However, statistics show that most of world's regions are impacted by (at least) some types of landslide phenomena that can be triggered by several factors including erosion, precipitation, earthquakes, human activity etc.

Landslides may occur rapidly or progress steadily at a fixed rate. They are common in soils and rockmasses with poor mechanical properties (highly fractured or weathered). However, a landslide can be triggered also by deformations along discontinuity layers of strong rocks. The nature and type of landslide phenomena are complex and are further analyzed below.

TYPES AND EXAMPLES OF SLOPE FAILURE

The most common and complete classification system of landslides is that provided by Varnes (1978), who introduces a system that requires the definition of the landslide material and the type of the movement induced. The ground materials are distinguished in five categories:

- ◉ **Rock:** An intact rockmass which used to be located at its initial position (i.e., it has not been eroded) before the movement occurred.

- ◉ **Soil:** Soil mass formed or transferred due to weathering and erosion of rocks. Soils consists of solid particles and voids filled with liquid and/or air, representing a three-phase system.

- ◉ **Earth:** The soil material in which more than 80% consist of particles smaller than 2 millimeters (upper limit of sand particles).

- ◉ **Mud:** The soil in which more than 80% consists of particles smaller than 0.06 millimeters (upper limit of silt particles).

- ◉ **Debris:** The soil material that has 20%-80% of particles that are bigger than two millimeters.

The five types of landslide movements that can be observed are categorized as following:

Falls

Falls are downward movements that progress rapidly and may not be preceded by initial movements or warnings. They occur when a rocky mass is detached from a slope along a discontinuity plane that can be associated with fractures, joints or bedding (Fig. 1). Falls are controlled by the shear strength of the discontinuity plane which is reduced with mechanical weathering propagation and the presence of water. Once detached, a rock boulder will follow a certain trajectory which depends on its size and shape and the topography of the region. The movement type can include free-fall, bouncing, rolling or a combination of those components.

Topples

Topples are also failures that occur in rocky materials and resemble falls. However, this type of failure is associated with a rotational movement that occurs around a certain point located in a relatively low position (Fig. 2). Topples are controlled by a combined action of gravitational forces that induce a bending moment and external forces (e.g., weathering, water pressure, freeze-thaw cycles).

Slides

Slides refer to ground movements along a specified surface or zone of weakness. A slide occurs when the shear stress applied along a surface overcomes its shear strength. The failure may propagate progressively initiating from a local failure zone. The main body of the slide will move downwards separating the stable from the unstable ground.

There are two main types of slides: **rotational** and **translational** slides. In rotational slides, the failure surface is curved inwards and points upwards and the landslide mass is approximately rotating around an axis transverse to the slide movement and parallel to the surface of the ground. The movement is usually associated with shear failure of the ground with its 3-dimensional geometry being "spoon-shaped". Certain features are defined to characterize a rotational slide as shown in Fig. 1.

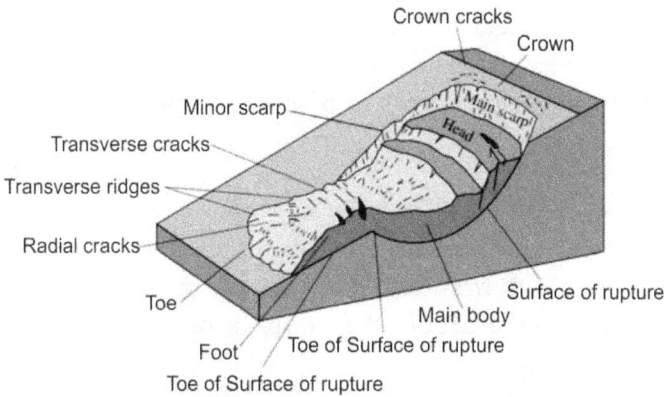

Fig. 1: Main features of a rotational landslide

The surface of rupture is the zone in which the ground material slides. The main scarp refers to a relatively steep edge at the head of the landslide that reveals the undisturbed ground and the visible component of the rupture surface. The crown is the area above the main scarp that has not moved downwards. The main body of the slide is the entire soil mass that has slipped along the failure surface. The head is the upper section of the slide between the main scarp and the displaced ground material. The toe is the most distanced part from the main scarp where landslide material has accumulated and the foot refers to the part of the failed material that has been deposited over the initial ground surface.

Nevertheless, sometimes the failure surface is controlled by pre-existing weakness planes (e.g., faults, fissures, cracks or joints). In this case, an engineering assessment must recognize those features since the failure would not be entirely controlled by the material's shearing component. A rotational slide will eventually stop propagating as a stress equilibrium is restored with the mass movement.

Lateral Spreads

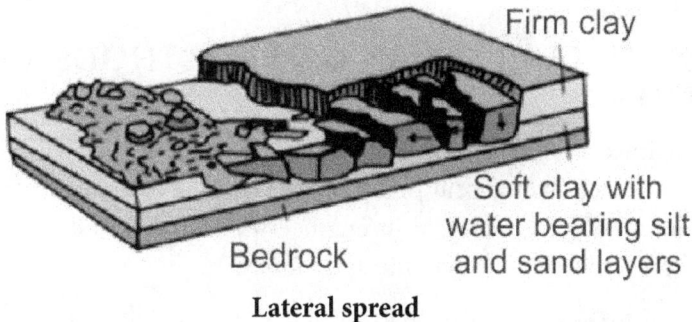

Lateral spread

Fig. 2. An illustration of a translational landslide

Lateral spreads are deformational phenomena caused by liquefaction, the process during which a saturated soil (usually sands) experiences loss of strength after a sudden change in its initial stress conditions. Therefore, the soil tends to behave more like a liquid than a solid. Such deformations occur on less steep slopes and are usually triggered by dynamic loads such as that of an earthquake. Lateral spreading is usually a progressive process that occurs mainly near shores, riverbanks, and ports where loose and saturated sandy soils exist. Infrastructure founded on those type of soils is prone to extensive damage (Fig. 2).

Flows

According to Varnes (1978), not all types of slope movements can be categorized in the aforementioned categories. There are certain landslide phenomena that take the form of slow or fast-moving flows. In rocks, there are types of slow movements that result in folding or bending. Due to the fact that these displacements resemble viscous fluids, they can be characterized as rock flows.

Regarding flows in soil materials, Varnes (1978) distinguished five main categories:

1. **Debris flow:** Quick soil mass movements that include less than 50% of fine material. Debris flows are usually triggered after heavy precipitation saturates and mobilizes the soil or they can be caused by another type of landslide upwards. They spatially occur near steep gullies where the conditions for such flows are favorable.

2. **Debris avalanche:** A rapid debris flow frequently triggered in steep slopes when the material's cohesion is relatively low or/and when the water content is high.

3. **Earthflow:** Flow in fine-grained soils or clayey rocks under saturated conditions. Liquefaction of the material leads to the formation of a bowl at the head of the slope and creates a unique "hourglass" effect.

4. **Mudflow:** A rapid flow consisting of at least 50% clay, silt and sand particles. They are characterized by a high-water content and sometimes they are referred to as mudslides.

5. **Creep:** A subtle and steady type of downward flow caused by shear deformation. Shear stresses are high enough to cause displacements but not adequate to result in shear failure. A common feature that reveals creeping flow is the presence of titled tree trunks.

CAUSES OF INSTABILITY, ANALISIS METHODS AND ASSUMPTIONS

A slope is an inclined boundary surface between air and the body of an earthwork such as highways, cut or fill, railway cut or fill, earth dams, levees and river training work. The slope stability analysis is crucial in engineering practice to ensure the stability of structures and prevent loss of human life and money.

The common methods for the analysis of a slope's stability are Culmann Method, Ordinary Method of Slices and Bishop Method of Slices. These methods are developed on the assumption that the plane of failure is circular arc, apart from the Culmann method that assumes a plane surface of failure through the toe of the slope.

The quantitative determination of the stability of slopes is necessary for a number of engineering projects, for instance, the design of earth dams and embankments, analysis of the stability of natural slopes, analysis of the stability of excavated slopes, and analysis of the deep-seated failure of foundations and retaining walls.

Causes of Slope Instability

1. Increased unit weight of soil by wetting.
2. Added external loads (moving loads, buildings etc).
3. Steepened slopes either by excavation or by erosion.
4. Shock loads.
5. Vibration and earthquakes.
6. Increase in moisture content.
7. Freezing and thawing action.
8. Increase in pore pressure.
9. Loss of cementing pressure.

Purpose of Slope Stability Analysis

1. Understand the development and form of natural slopes and the processes responsible for different natural features.
2. Assess the stability of slopes under short-term (often during construction) and long-term conditions.
3. Evaluate the possibility of landslides involving natural or existing engineered slopes.
4. Analyze landslides and understand failure mechanisms and influence of environmental factors.
5. To redesign failed slopes and plan for the design of preventive and remedial measures, where necessary.
6. Study the effect of seismic loading on slopes and embankments.

TYPES OF SLOPE FAILURE

1. **Circular slips:** They are related to homogenous, isotropic soil conditions.
2. **Non-circular slips:** Non-circular slips are associated with non-homogenous soil conditions.
3. **Translational failure:** This type of failure takes place where the form of failure surface is affected by the presence of an adjacent stratum of different strength, and the adjacent stratum is fairly shallow.
4. **Compound Failure:** It occurs where the form of failure surface is affected by the presence of an adjacent stratum of different strength, and the adjacent stratum is relatively deep.

SLOPE STABILITY ANALYSIS ASSUMPTIONS

1. Problems are two dimensional.
2. Coulomb's theory can be used to compute shear strength.
3. Shear strength is assumed as uniform along the slip surface.
4. The flow net in case of seepage can be drawn and seepage forces evaluated.

Factor of Safety

There are different safety factors which are used in the analysis of slope stability. For instance, factor of safety with respect to strength, cohesion, friction, and height. the former is widely used.

1. Safety Factor with Respect to Strength: It is the ratio of the maximum load or stress that a soil can sustain to the actual load or stress that is applied, and expressed as follows:

$$F = \frac{\tau_{ff}}{\tau} \qquad (1)$$

Equation 1 can be modified slightly and expressed as in Eq. 2.

$$\tau = \frac{c}{F} + \frac{\sigma_n \tan\theta}{F} \qquad (2)$$

Where

τ = Actual shear stress applied to the soil.

τ_{ff} = Maximum shear stress that a soil can sustain at the value of normal stress σ_n.

2. Safety Factor with Respect to Cohesion: It is the ratio between the actual cohesion and the cohesion required for stability when the frictional component of strength is fully mobilized, expressed using the following formula:

$$\tau = \frac{c}{F_c} + \sigma_n \tan \theta \qquad (3)$$

3. Safety Factor with Respect to Friction: It is the ratio of the tangent of the angle of shearing resistance of the soil to the tangent of the mobilized angle of shearing resistance of the soil when the cohesive component of strength is fully mobilized.

$$\tau - c + \frac{\sigma_n \tan \theta}{F_c} \qquad (4)$$

4. Safety Factor with Respect to Height: It is the ratio between the maximum height of a slope to the actual height of a slope.

$$FH = \frac{H_{max}}{H} \qquad (5)$$

SLOPE STABILITY ANALYSIS METHODS

1. **Culmann Method:** It is not widely used because it is demonstrated that plane surfaces of sliding are noted only with very steep slopes, and for relatively flat slopes the surfaces of sliding are almost always curved.

2. **The Zero angle of Shearing Resistance Method:** This method of slope stability analysis is based on the assumption that the plane of failure is in the form of a circular arc. It is a practical method for the evaluation of the short-term stability of saturated clay slopes.

3. **Ordinary Method of Slices:** It is considered where the effective angle of shearing resistance is not constant over the failure surface, such as in zoned earth dams where the failure surface might pass through several different materials.

4. **Bishop Method of Slices:** In Bishop method of slices, the analysis is conducted in terms of stresses rather than forces which are used in the ordinary method of slices. The main difference this method and the Ordinary Method of Slices is that resolution of forces takes place in the vertical direction instead of a direction normal to the arc. The simplified Bishop method of slices provides a safety factor which is considerably close to those evaluated using more rigorous methods of analysis.

MECHANICS OF UNSATURATED SOILS FOR AGRICULTURAL APPLICATIONS

Introduction

Soils undergo intensive changes in their physical, chemical, and biological properties during natural soil development and as a result of anthropogenic processes such as plowing, sealing, erosion by wind and water, amelioration, excavation, and reclamation of devastated land. In agriculture, soil compaction as well as soil erosion by wind and water are classified as the most harmful processes.

Irrespective of land-use systems, soil deformation as the sum of soil compaction and shear processes leads to numerous environmental changes affecting soil quality for crop production, soil biodiversity, filtering and buffering functions of soils, soil-water household components, trace gas emissions, soil erosion and nutrient export, and related off-site effects.

In forestry, normal plant and soil management, tree harvesting, and clear-cutting also affect site-specific properties, including organic-matter loss and groundwater pollution and gas emissions, which have the potential to cause global changes. Furthermore, soil amelioration especially by deep tillage prior to replanting often causes irreversible changes in properties and functions. Oldeman (1992) showed that about 33 million ha of arable land are already completely devastated by soil compaction in Europe alone while the total area of degraded land worldwide exceeds about 2 billion ha. Physical (soil erosion and deformation) and chemical processes are responsible for about 1.6 and 0.4 billion ha of degraded soils, respectively. Worldwide population growth will reduce the average area per person for food and fiber production from 0.27 ha today to <0.14 ha within 40 years, and even if the advances in technological developments continue to grow, more concern has to be made in order to prepare enough food for the population worldwide.

Consequently, a more detailed analysis of soil and site properties is needed to manage soils in accordance with their potential properties. Numerous attempts to predict the strength of the soils were made; however, all are based on static loading experiments, which assume that equilibrium in settlement for stepwise increasing loads is achieved. This approach neglects the dynamic nature of soil loading processes during field traffic where loads are applied mostly short term and where multiple loading events may cumulatively add to the total compression. Thus, the increase in irreversible deformation of arable soil by agricultural machinery can not only be related to the increasing mass of machines (defined as static approach) but also enhanced by the increase of wheeling frequency as a dynamic component. The relevance of wheeling frequency can be estimated based on the calculations of Olfe (1995). Considering an average-sized wheat-production farm, the number of load repetitions in a time period of 5 years may add up to 50 events for 85% of the field and up to 100 events for permanent wheeling tracks.

An additional threat concerning soil degradation is caused by tillage erosion as the actual soil loss can exceed by far 20 Mg ha^{-1} and per tillage operation. This effect is the more pronounced, the drier the soil, the weaker the aggregates, the higher the dynamic energy input by agricultural machinery, and the greater the field size. The kinetic energy applied during seedbed preparation under dry conditions alone already results in an increased mass transport by wind. Additionally, the unproductive water loss due to evaporation from an increased accessible pore and particle surface and the enhanced organic-matter decomposition due to tillage both contribute to global change problems.

Soil creep, mudflow after rainstorms, and intensified surface-water runoff are due to reduced/prevented vertical water infiltration in soils generating pronounced lateral fluxes of water, solids, and nutrients. The loss of shear strength, defined by the angle of internal friction and cohesion, of near-saturated soils subject to buoyancy forces that cause such lateral fluxes produces tremendous damage to the landscape and human beings. Related economic impacts are estimated to exceed billions of euros worldwid.

In addition to productivity losses of soils in farming regions, the loss in biodiversity and the effects on global changes by modifying physicochemical processes result in the emission of greenhouse gases and the mobilization of heavy metals in soils due to redox reactions, which in turn cause groundwater pollution threats.

If, on the other hand, the tilled soil dries out, transport by wind may occur, resulting in severe reductions in potential site productivity. Furthermore, the preparation of a seedbed leads to abrupt changes in the transport of gas, water, ions, and heat between the tilled and deeper soil layers. This is especially true concerning preferential flow through structured soils.

When it comes to the quantification of soil deformation with respect to the induced changes of physical functions, again the dynamic nature of soil loading processes during field traffic must be considered, where loads are applied mostly short term and where multiple loading events may cumulatively add to the total compression, although the statically determined precompression stresses are not exceeded and should result in no further compression.

Such anthropogenic changes make clear that the discussion of soil process dynamics within the soil profile is most relevant. Some of the interactions between soil structure, water status, and aeration of structured soils in relation to root growth and compressibility of arable land have been described by Emerson et al. (1978). In current discussions on sustainability, soil compaction is repeatedly mentioned as one of the main threats to agriculture, which should be avoided.

The aim of this chapter is to analyze the relationships between stress, strain, and strength as well as the consequences of soil deformation for physical and physicochemical properties. It also includes a description of well-established and new methods for measuring mechanical properties of soils and introduces possibilities for modeling soil strength and stress–strain relationships from the micro- to the macroscale.

Stress–Strain Relationships in Soils

Stress Theory

Definitions

Before discussing the methods of field and laboratory stress measurement as well as factors influencing compaction, one needs to differentiate between several terms used to define compressive properties.

Mechanical Stresses

Stress is defined as force per area within a solid body. Stress can be induced by external or internal forces, which lead, if the body is nonrigid, to a change in the body's volume and/or shape expressed as deformations or strains. The mechanical behavior of a soil can therefore be characterized by its stress–strain relationships.

Strength is typically referred to as the maximum amount of stress a solid material can withstand before it fails; thus, exceeding soil strength results in soil failure or yield.

Strength depends on internal parameters such as particle-size distribution, type of clay minerals, nature and amount of adsorbed cations, content and type of organic matter, aggregation induced by swelling and shrinkage, stabilization by roots and humic substances, bulk density, pore-size distribution and pore continuity of the bulk soil and single aggregates, water content, and/or matric potential.

Stress State

The number of stress state variables required to define the stress state depends primarily upon the number of phases involved. The effective stress $\acute{\sigma}$ can be defined as a stress variable for saturated conditions and is the difference between the total stress (σ) and the neutral stress (u_w), which is equal to the pore water pressure:

$\sigma' = \sigma - uw,$

where

◎ σ' is transmitted by the solid phase

◎ u_w is transmitted by the liquid phase, respectively

For silty and clayey soils, the values of the parameters in Equations 3.1 and 3.2 depend on soil aggregation, pore arrangement and strength, and hydraulic properties. Thus, the material function of the components in structured soils is only valid as long as the internal soil strength is not exceeded by the externally applied stresses. It changes if, for example, aggregates are destroyed during soil deformation and the structural properties reduce to those that depend merely on texture.

The extent of soil deformation can be described by stress–strain relationships and by their relative proportions. In the absence of gravitational and other applied forces, effective stresses in three-phase soil systems can be expressed as a tensor containing three normal stresses and six shearing components; assuming symmetry of the stress tensor, the shear stresses are reduced to three independent components. Therefore, three normal stress ($\sigma_x, \sigma_y, \sigma_z$) and three shear stress ($\sigma_{xy}, \sigma_{xz}, \sigma_{yz}$) components must be determined to fully define the stress state at every location in a solid body.

A graphical representation of the stress state variables and the action of water menisci at intergranular contacts in an unsaturated soil is provided in Figure 3.

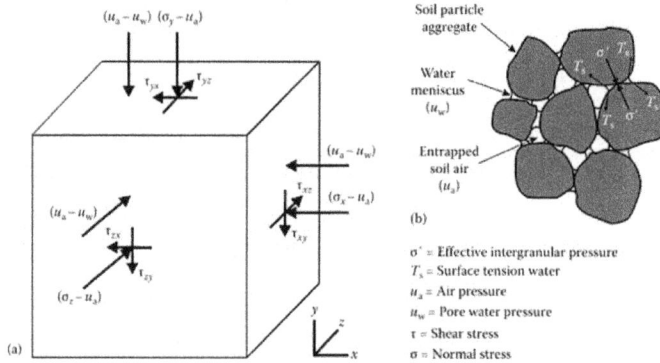

Fig. 3 **(a) Stress state variables for an unsaturated soil. (b) Schematic sketch of water menisci at interparticle/interaggregate contacts illustrating the effect of pore water pressure (u w) on intergranular tension stresses.**

Stress Propagation

Each applied force per contact area is transmitted into the soil in three dimensions and can alter the physical, chemical, and biological properties of the soil (e.g., water infiltration, rootability) if the internal mechanical strength is exceeded. Stress propagation theories are rather old and have been often modified and adapted to in situ situations. The fundamental theory of Boussinesq (1885) is only valid for completely elastic material, while Fröhlich (1934) and Soehne (1958) included elastoplastic properties through the introduction of concentration factor values (u_k). More comprehensive descriptions of these models are given by Koolen and Kuipers (1983), Bailey et al. (1986), Johnson and Bailey (1990), and Bailey et al. (1992). Horn et al. (1989) introduced precompression stress (r_v)-dependent values for the concentration factor, which are smaller in the recompression stress range, while they increase in the virgin compression stress range. The latter can be explained by the plastic deformation behavior, which causes a deeper stress transmission closer to the perpendicular line.

Hydraulic Stresses

The effective stress defines the forces per given area, which can stabilize the soil particles against any kind of soil deformation. The hydraulic stress component (u_w) can either be negative (concave water menisci) or positive (convex water menisci). In case of convex menisci, it can result in weakening of the total soil system, especially when shear forces are applied. In case of pure compression under saturated conditions, however, effective and hence compressive stresses are reduced while the neutral stress (water pressure) bears part or even the entire externally applied load. In most cases, air pressure (u_a) is ignored, assuming that gas pressure in soil during loading is in equilibrium with atmospheric conditions.

If, on the other hand, hydraulic stresses become negative (=suction), the contractive forces even increase the effective stress. In addition to capillary forces associated with water, salt effects (i.e., water potential = sum of matric and osmotic potential) and hydrophobic substances can also increase soil strength by altering wetting angle, cohesion, internal friction, as well as elastic and viscous properties of the soil.

Hydraulic stresses result in shrinkage in almost all soils. In an initially homogenized state, soils undergo proportional and thereafter residual shrinkage during the first drying phase. Depending on the history of the formerly applied hydraulic and mechanical stresses, the shrinkage curve pattern, however, shifts and it must be now differentiated between structural shrinkage (=structural rigidity), virgin, and residual shrinkage behavior. Structural shrinkage defines the internal soil strength caused by deformation related to capillary forces. Structural shrinkage is more pronounced, the more often and intensive soil dries out. Horn (1994a), for example, described the effect of the drying frequency and intensity on shrinkage behavior and pointed out that soil strength increases the more often and longer contracting water menisci occur while only a few but very intense drying processes are mostly not capable of rearranging particles to form a rigid structure with smaller entropy.

The power of capillary forces and their strengthening potential for soil architectures can be demonstrated by a simple computation. Given the mass of the earth ($\sim 6 \times 10^{24}$ kg) and the area of, for example, Germany ($\sim$360,000 km^2) of which 50% is arable soil (=180,000 km^2) and assuming a thickness of the plowed topsoil of 30 cm with a hypothetical specific surface area of 800 m^2 g^{-1} (for simplicity we assume the soil to consist of smectitic clay and %4–%2 of organic carbon), we would end up with a total surface area of ca. 10×6.5^{19} m^2. If we furthermore presume that water menisci at a potential of pF 4 would be effective over the total surface (c→ 1, assuming a very high negative air-entry pressure), then the sum of menisci forces acting within the topmost 30 cm of arable soils in Germany could carry the total mass of the earth.

Stress Coupling

Based on the findings by Baumgartl (2003), who described the similarity of mechanical stress–strain curves and shrinkage curves resulting from hydraulic stresses, reserachers proved that the link between mechanical and hydraulic "pre"stresses results in nearly identical changes in shrinkage curves. Additionally, they explained the differences between mechanically induced collapse and matric potential-dependent shrinkage patterns by the dimensionless ÷ factor, which, however, is difficult to determine experimentally. Accounting for the interactions between mechanical and hydraulic processes is very important to understand the deformation behavior of unsaturated soils. Nevertheless, their description is inherently complex involving effective stress theory, loading rate dependency, and nonlinear hysteretic coupled transport and stress–strain functions.

According to Richards and Peth (2009), mechanical and hydraulic processes can be coupled by treating the initially independent load-deformation and transport processes by an incremental approach as shown in Figure 4.

This coupling is most important in modeling the interaction between water flow and deformation/failure as, for example, encountered in swelling soils, landslides on slopes during rainstorms, and consolidation of high water content soils. In a stress-coupling routine, changes in pore water pressure (h) are calculated as a function of changes in stress (σ) or strain (ε), and in a moisture-coupling routine, changes in stress or strain are calculated as a result of pore water pressure changes related to both water transport and load deformation.

A detailed mathematical description of the finite element model (FEM) used for mechanical and hydraulic stress coupling and the underlying constitutive relations and material parameters.

Soil Strength Changes due to Stress-Dependent Effects on Hydraulic Properties

Total stress application affects interparticle bonding as well as pore-size distribution, pore geometry, and the degree of saturation, which either weakens the soil during compression or makes it even stronger. While the former is caused by soil settlement, reducing pore volume and hence increasing the degree of saturation, the latter is related to the stress-dependent water redistribution in newly formed finer pores at the expense of bigger and already air-filled ones causing more negative matric potential values, thus increasing effective stresses. Researchers describe the effect of static loading on the transition between stress-induced soil strengthening and weakening and found that this change occurs close to the precompression stress value.

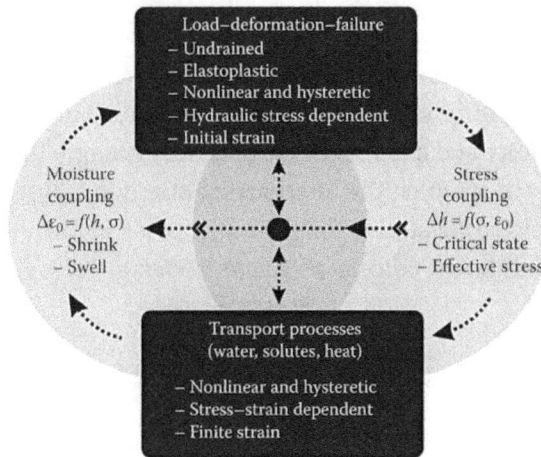

Fig. 4 Conceptual framework for modeling coupled soil mechanical and hydraulic processes. The stress-coupling routine models pore water pressure changes (Dh) as a function of stress (s) or strain (e0). The moisture-coupling routine models strain changes (De0) as a function of pore water pressure (h) and effective stress (s$_2$).

Soil Strength Changes due to Stress-Dependent Effects on Hydraulic Properties

The changes resulted initially in a strength increase and the pore water pressure became more negative, while at higher stresses the pore water pressure even reached positive values starting from "30 kPa initial matric potential. If external forces are applied as repeated short-time loading (e.g., due to wheeling or trampling) or if shear deformation-induced rearrangement of particles occurs, we also have to consider the "water menisci pumping effect due to short-term suction changes" because it enhances the weakening process due to a more pronounced and more complete swelling, finally resulting in a complete soil homogenization. Researchers measured stress-induced changes in pore water pressure in paddy soil samples. Increasing frequency of loading and unloading changed water menisci forces from negative to positive pore water pressure values and caused complete soil softening. Similar effects were also reported for short-term cyclic loading of homogenized soil, where even when the soil was unsaturated at an initial matric potential of "6 kPa, positive pore water pressure was measured during the short-term loading phases. Water-induced soil softening is also evident when shear forces are applied to the soil surface, for example, by wheeling due to slip effects.

Researchers proved the effect of shear-induced changes in pore water pressure in comparison with the previously applied static stress (Fig. 5). Even if the consolidated stress–strain state had been reached due to a given static loading, the following so-called consolidated shear process again caused an additional compaction and an extra change in the pore water pressure values. The more rapid the shear process occurs and the smaller the hydraulic conductivity even at increased hydraulic gradients, the more positive become the pore water pressure values and the weaker the total soil.

Shear Deformation

Shearing a soil volume always leads to a pronounced volume constant rearrangement of particles often resulting in significant changes in pore functions by decreasing pore connectivity or continuity and changes in interparticle strength due to a loss of cohesion (Fig. 6). The stress/strain at which the interparticle strength is lost and the soil begins to fail is referred to as yield stress/strain. The resistance of a soil toward shear deformation is described by shear moduli (G), (*rheometrical parameters*).

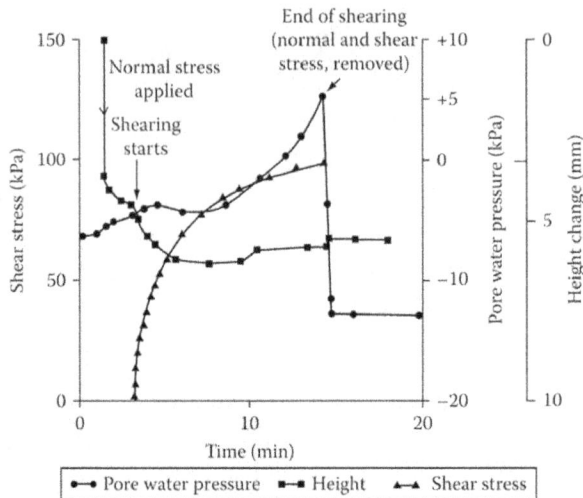

Fig 5 Changes in shear stress, sample height, and pore water pressure during shearing. The samples were predried to -6 kPa and pre-consolidated prior shearing. Upon shearing at the same applied stress, the sample further consolidated (height change) and pore water pressure increased reaching even positive values during shearing. After normal and shear stress were removed, pore water pressure decreased to a final value more negative than the initial pore water pressure.

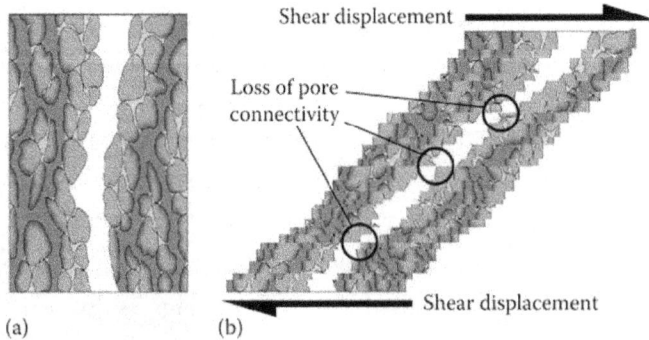

Fig. 6 Schematic sketch of the effect of shear displacement on pore continuity (b) in comparison with an "unstressed" pore system (a).

Liquid Limit

The liquid limit is determined with a *Casagrande* apparatus (Casagrande, 1932) where a homogenized moist sample is placed in a special bowl and a V-shaped trench is cut from the top to the bottom. The bowl with the sample is bounced up and down by a special crank onto a hard rubber block at a frequency of two strokes per second until the trench is closed at a length of <1 cm. The number of strokes when this condition is met is counted and the gravimetric water content determined on a subsample at the end of the test. The liquid limit is defined as the water content ($\dot{E}_g$) where the test criteria are fulfilled at 25 strokes corresponding to a certain "standardized" amount of energy applied to the soil. The test is repeated for at least four water content values and the number of strokes plotted versus the water content on a semilogarithmic paper, finally producing a linear line. The liquid limit increases with clay and organic-matter contents, ionic strength, cation valency, and proportion of 2:1 type clay minerals in the soil.

lastic Limit

The plastic limit is defined as the water content ($\dot{E}_g$) when homogenized soil samples begin to crack and crumble when rolled to a diameter <3 mm. It has practical implication in civil engineering and agriculture since it determines the lower critical water content at which soils become plastic and hence difficult to till or excavate.

Plasticity Index

The difference between the water contents (Θ_g) at the liquid limit (LL) and plastic limit (PL) is the so-called plasticity index (PI = LL – PL), which is often used as an indicator for soil workability. With an increasing plasticity index, soils become more sensitive to plastic deformation. The higher the plasticity index the smaller is the water-content range at which soils can be tilled efficiently and trafficked without considerable soil deformation. The smaller is also the angle of internal friction over a wider range of water contents. Many attempts have been made to correlate PI to soil strength. In principle, this test only gives information on minimum strength values correlated with soil water content. It is furthermore limited to silty, loamy, and clayey soil samples while such correlation fails and is too insensitive for sandy material. Thus, especially coarse-textured soils cannot be classified by these data.

Rheometrical Parameters

Advances in rheological research as an independent science have brought sophisticated techniques for quantifying flow behavior and deformation properties of viscoelastic substances (mainly fluids)—the most important parameters are shear modulus (G), plastic viscosity (η_p), and yield stress ($\sigma\tau_y$).

Although first rotational viscometers (1888) and later rheometers (~1951) have been developed, early modern microprocessor-based systems, which are available since the 1980s, added more options for rheological testing including creep, relaxation, and oscillation tests. The methods are today well established and widely used in polymer, chemical, and material sciences as well as in the paper and food industries.

Despite its potential applicability, however, they are still not commonly employed in soil science research. One of the earliest contributions has been made though by Yasutomi and Sudo (1967) who used a low-frequency forced oscillation viscometer to study viscoelastic properties of soil. Later, Keren (1988) and Hesterberg and Page (1993) used a viscometer to investigate the influence of electrolyte composition and pH on rheological characteristics of clay mineral suspensions. Studies based on mineral suspensions certainly aid the understanding of the principle influence of surface chemistry on mechanical solid–interface interactions but have the drawback of ignoring the pertinent action of water as a lubricant during deformation processes.

More recently, researchers have used a rotational rheometer with a parallel-plate sensor system to determine these *rheometrical parameters* on natural soils and clay samples under more realistic field-moisture conditions. Higher water contents and high stress amplitudes resulted in lower viscosity of all investigated soils. Although this is not surprising, the authors could demonstrate that rheometry is a suitable and sensitive technique that enables us to quantify rheological properties of soils based on physically well-defined boundary conditions. Also Markgraf et al. (2006) investigated rheological characteristics of natural soil samples (Avdat loess, smectitic vertisol, and kaolinitic oxisol) using a parallel-plate rheometer. By conducting so-called amplitude sweep tests (oscillatory test with continuously increasing deformation), they were able to reveal differences in structural stabilities between the smectitic vertisol and the kaolinitic oxisol and the influence of electrolyte concentrations on rheological parameters of Avdat loess samples.

The sample (usually a remolded paste at defined water content) is placed between two parallel plates (diameter 2.5 cm) with an adjustable gap size commonly between 2 and 4 mm. To set up the test, the upper plate is slightly pressed against the sample to ensure contact and time is given for stress dissipation. To avoid slip at the interface between the sample and the plates, both the upper and lower plates are furrowed. In an amplitude sweep test, the upper plate rotates (oscillatory) at a predefined frequency (e.g., 0.5 Hz) and variable amplitude beginning with very small shear deformations (corresponding to angular deflection) up to the maximum shear deformation where the angular deflection is equal to the gap spacing between the plates (e.g., 4 mm deflection at a gap spacing of 4 mm). Shear strain ã is expressed as percentage ranging from 0% (no deflection) to 100% (maximum deflection).

Mechanical Properties of Homogenized Soils

If only mechanical properties of homogenized soils are compared, gravelly soils are less compressible than sandy or finer-textured soils, and any deviation of the particles from the spherical shape results in an increase in the shearing resistance and soil strength. Researchers found that the angle of internal friction for medium and coarse sands at comparable bulk density and water content but of different origins (quartz, basalt, volcanic ash, and shells) ranged from 25° to 50°. Soil strength also depends on the type and content of clay minerals and exchangeable cations. At the same bulk density and pore water pressure, compressibility of homogenized soil increases with clay content and decreases with soil organic-matter (SOM) content. *Vice versa* soils at the same clay content are more readily compressed at lower bulk density or higher moisture content. Cohesive forces between illite, smectite, and vermiculite are greater than for kaolinite, which in turn has a larger angle of internal friction because of its size and particle shape. Additionally, the higher the valency of the adsorbed cations and the higher the ionic strength of the soil solution, the greater is the shearing resistance under otherwise similar conditions.

Shearing resistance also increases depending on both the nature and content of SOM. In agricultural soils with the same physical and chemical properties, the mechanical parameters (precompression stress, angle of internal friction, and cohesion) are greater when the relative contents of carbohydrates, condensed lignin subunits, bound fatty acids, and aliphatic polymers are higher.

8

GEO-ENVIRONMENTAL ENGINEERING

Geo-environmental Engineering is the engineering of *geologic* (earthen) and *Geo-synthetic* (polymer) materials for problems related to the protection of human health and the environment. The primary problems addressed by Geo-environmental Engineers pertain to the protection of uncontaminated regions of the subsurface as well as the remediation or *clean up* of regions of the subsurface that have been contaminated by one or more events (e.g., industrial chemical spills, leaking waste containment facilities, leaking above-ground and underground storage tanks, infiltration of pesticides, etc.). Since the nature of the problems addressed in Geo-environmental Engineering is diverse, solutions to *Geo-environmental problems* typically require the expertise of a variety of professionals who possess a similar diversity in terms of educational background and training. Because of this diversity, efficient and effective technical interaction among these professionals can be problematic. Thus, professionals who have attained a breadth of knowledge in a variety of the disciplines associated with Geo-environmental problems can facilitate the professional interaction needed for successful completion of Geo-environmental projects within a *multidisciplinary setting*.

Geo-environmental engineering is an emerging and exciting field that offers numerous technical challenges and great opportunities to understand multi-disciplinary problems and develop solutions to protect public health and the environment and encourage sustainable development.

Geoenvironmental Engineering includes issues such as the following:

- ◉ Effects of waste liquids on barrier material properties (compatibility)
- ◉ Containment strategies for emerging waste forms (e.g. nano-waste, pharmaceutical waste, animal waste (prions))
- ◉ Contaminant transport through low-permeability soil barriers
- ◉ Development and evaluation of novel containment barrier materials

- Development and evaluation of subsurface remediation technologies
- Diffusion through polymer materials used as components of engineered barrier systems
- Evaluation of flow and transport models for predicting barrier and remediation system performance
- Geosynthetics for containment and remediation applications
- Geotechnical aspects of waste containment and remediation systems
- Leaching of contaminants from stabilized/solidified waste forms
- Mine waste containment and remediation
- Physical, chemical, and biological processes governing in situ remediation technologies
- Physico-chemical interactions between soils and contaminant liquids (sorption, ion exchange)
- Unsaturated flow through soil covers used for waste disposal

SCOPE OF GEOENVIRONMENTAL ENGINEERING

- Any project that deals with the interrelationship among environment, ground surface and subsurface (soil, rock and groundwater) falls under the purview of geoenvironmental engineering.
- Scope is vast and requires the knowledge of different branches of engineering and science put together to solve the multi-disciplinary problems.
- Engineer should work in an open domain of knowledge and should be willing to use any concepts of engineering and science to effectively solve the problem at hand.

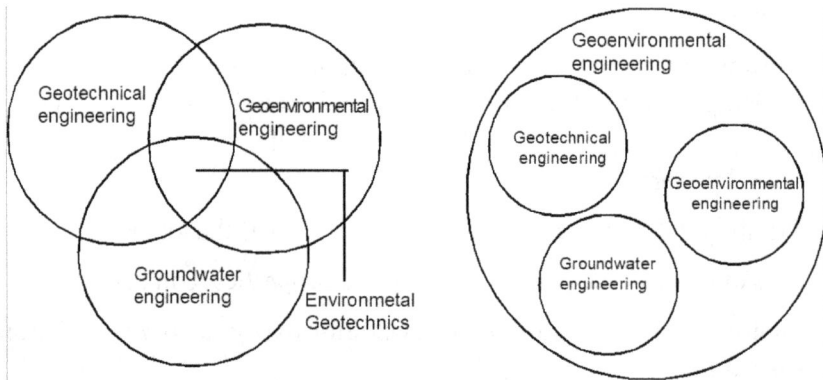

Fig. 1

- Most challenging aspect is to identify the unconventional nature of the problem, which may have its bearing on multiple factors, e.g. an underground pipe leakage may not be due to the faulty construction of the pipe but caused due to the highly corrosive soil surrounding it. Corrosiveness may be attributed to single or multiple manmade factors, which need to be clearly identified for the holistic solution of the problem.

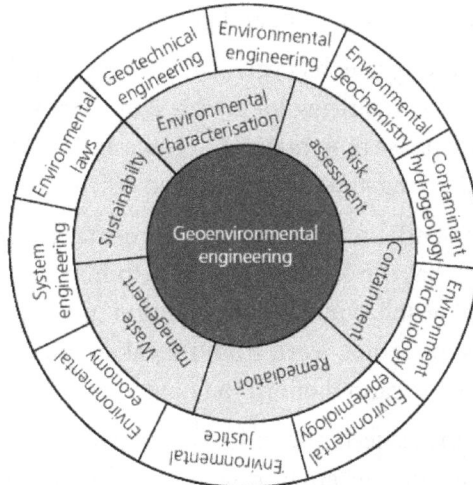

Fig. 2

◉ The conventional approach of assessing the material strength of the pipe alone will not solve the problem at hand.

◉ For achieving the "Green Environment". Despite a lot of effort, it is very difficult to cut off the harmful effects of pollutants disposed off into the geoenvironment.

◉ The damage has already been done to the subsurface and ground water resources, which is precious.

◉ An effective waste containment system is one of the solutions to this problem. However, such a project has different socio-economic and technical perspectives.

◉ Realization of such projects require the contribution of

    a. environmentalist,

    b. Remote sensing experts,

    c. Decision makers,

    d. Common public during its planning stage,

    e. Hydrologists,

    f. Geotechnical engineers for its execution stage and

    g. Several experts for management and monitoring of the project

◉ The totality of the problem can be visualized under the umbrella of geoenvironmental engineering.

◉ The real challenge for a geoenvironmental engineer is how well he can integrate the multi-disciplinary knowledge for achieving an efficient waste containment.

- In most parts of the world, damage has already been done to the geoenvironment and groundwater reserves due to indiscriminate disposal of industrial and other hazardous wastes.

- Owing to excessive demand, it becomes important to remediate and revive the already polluted geoenvironment and groundwater.

- Geoenvironmental engineer has a great role to play for deciding the scheme of such remediation practice, requiring lot of concepts from soil physics, soil chemistry, soil biology, multi-phase flow, material science and mathematical modeling, need to be taken for planning and execution of an efficient remediation strategy.

- Therefore, it is essential for the geoenvironmental engineer to think out of the box, to an extent that the knowledge can help him visualize the problem better and suggest efficient solution. Else, the solution to such problems becomes a trial and error process or rather, learn from mistakes and rectify. Since such projects are cost intensive one cannot afford to take too much of chances.

- Another important issue is the reuse and recycling of waste materials, which reduces the burden on our environment manifold, e.g. example is exploring the possibility of mass utilization of fly ash for geotechnical applications.

- For using the waste meaningfully, there are issues like short term and long term impact of such process, which is a governing factor for deciding its selection as a viable material.

- Geoenvironmental engineering is to simplify the process of understanding the behavior and resort to reliable predictions and estimations. This would require a thorough knowledge on material science and chemistry and the reaction it undergoes with time. This is indeed a tough task, but needless to say, such challenges make this subject quite interesting

- Examples (i) frequent occurrence of landslides, and flood protection works.

- Geoenvironmental engineering is more research oriented and new concepts and methodologies are still being developed.

MULTIPHASE BEHAVIOR OF SOIL

- Classical soil mechanics assumes soil media to be completely water or air saturated, which is a typical example of a two phase media consisting of soil solids and water/air. The assumption of two phases considerably simplifies the mathematical quantification of the complex phenomena that take place in porous media.

- Of late, geotechnical and geoenvironmental engineering problems require the concept of three or multiphase behaviour of soil for realistic solution of several field situations. e.g. a partially saturated soil is a three phase porous media consisting of air, water and soil. The three phases result in transient and complex behaviour of unsaturated soil.

⊙ Such cases are encountered while designing waste containment facility where flow characteristics of unsaturated soil need to be determined. When it comes to soil-water-contaminant interaction there are multi-phase interactions involved.

⊙ The migration of non-aqueous phase liquid (denoted as NAPL) through porous media is a typical example. Fluidized bed, debris flow, slurry flow, gas permeation through unsaturated soil media are some problems where multiphase behaviour becomes important.

⊙ Such studies are handy while designing remediation scheme for contaminated soil and groundwater, which are very important issues for the geoenvironmental engineer to solve.

ROLE OF SOIL IN GEOENVIRONMENTAL APPLICATIONS

⊙ All civil engineering structures are ultimately founded on soil and hence its stability depends on the geotechnical properties of soil.

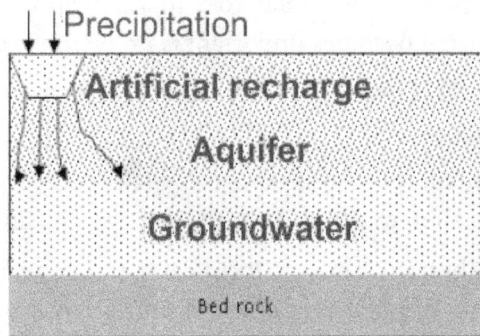

Fig. 3 Artificial groundwater recharge

⊙ Conventional geotechnology is more concerned about rendering soil as an efficient load bearing stratum and designing foundations that can transfer load efficiently to subsurface.

⊙ Apart from this, soil is directly related to a number of environmental problems, where the approach should be a bit different.

⊙ Consider the case of groundwater recharge as shown in Fig. 3.

⊙ The infiltration and permeation property of homogenous or layered soil mass above water table decides the rate of recharge.

⊙ In this case, a geotechnical engineer has to work closely with hydrogeologists for deciding different schemes of artificial groundwater recharge.

⊙ Consider the case of waste dumped on ground surface.

⊙ During precipitation, water interacts with these wastes and flow out as leachate. When the leachate flows down, soil act as buffer in retaining or delaying several harmful contaminants from reaching groundwater.

⊙ Such a buffering action obviously depends on the texture and constituents of soil mass.

⊙ While designing a waste containment facility, the role of soil in such projects is enormous. A coarse grained soil with filter property is required for leachate collection where as a fine grained soil is required for minimizing flow of leachate.

⊙ These are two entirely different functions expected from soil in the same project. The cap provided for waste dumps also necessitate the use of specific type of soils with the required properties.

⊙ Special type of high swelling soils is used as backfills for storing high level radioactive waste in deep geological repositories. Another important geoenvironmental problem, namely, carbon sequestration uses the geological storage capacity for disposal of anthropogenic CO_2 to mitigate the global warming.

⊙ Therefore, soil plays a very vital role in geoenvironmental projects and the property by which it becomes important is problem-specific.

⊙ Importance of Soil Physics, Soil Chemistry, Hydrogeology and Biological Process

⊙ Soil physics is the study of the physical properties and physical processes occurring in soil and its relation to agriculture, engineering and environment.

⊙ It deals with physical, physico-chemical and physico-biological relationship among solid, liquid and gaseous phase of soil as they are affected by temperature, pressure and other forms of energy.

⊙ The concepts of soil physics is used for determining the transport of water, solute and heat (matter and energy) through porous media, which is important to solve the problems related to subsurface hydrology, groundwater pollution, water retention characteristics of soil, improving crop production, rainfall induced landslides etc. Soil physics is mostly quantitative and mathematical in nature and requires the knowledge of soil physical properties.

⊙ The emergence of discipline "soil chemistry" began when J. T. Way (father of soil chemistry) realized that soil could retain cations such as NH_4^+, K^+ in exchange for equivalent amounts of Ca^{+2}.

⊙ Knowledge of soil chemistry is important to understand interactions between soil solids, precipitates and pore water, including ion exchange, adsorption, weathering, buffering, soil colloidal behaviour, acidic and basic soils, salinity etc.

- Understanding subsurface for geoenvironmental problems requires extensive knowledge of hydrogeology. Hydrogeologic parameters influence a lot on how a waste containment facility performs over its design life.

- In biological processes occurring in soils some type of microorganisms such as Pseudomonas aeruginosa is used for remediation of hydrocarbon contaminated site. It is very essential to understand the rate of such reaction and the impact of such remediation.

SOURCES AND TYPE OF GROUND CONTAMINATION

- Solid, liquid and gaseous waste forms contaminates subsurface and groundwater due to indiscriminate disposal. Solid wastes come from municipal, domestic and industrial sources.

- Municipal wastes amounts to around 50 percent of the total wastes produced. Household, hospital, agricultural wastes forms part of municipal wastes. Returning these wastes to soil is considered to be a low cost option.

- Abandoned e-waste, batteries, vehicles, furniture, debris from construction industry is considered as solid waste and is produced from both urban and rural areas.

- Large scale industrial development produces huge quantities of hazardous waste and the sources are iron and steel industries, packaging factories, paints, dyes, chemicals, glass factories, fertilizer and pesticide industries, mine excavation waste etc. Coal mining, radioactive fuel mining, petroleum mining and thermal power plants generate hazardous solid waste that requires effective management.

- The main source and type of hazardous liquid waste include industrial waste water contained in surface impoundments, lagoons or pits. It is also produced from municipal solid refuse and sludge that are disposed on land.

- If not handled properly sewage becomes an important source of liquid waste that has undesirable effect on environment. Petroleum exploration leaves waste brine solution which needs to be managed to prevent groundwater pollution. Liquid waste emerges due to mining operation which is hazardous.

- Some of the gaseous waste includes NOx, CO, SO2, volatile hydrocarbons etc. Chemical reaction may take place in air producing secondary pollutants. SO_2 combines with oxygen to produce SO3, which in turn combines with suspended water droplets to produce H2SO4 and fall on ground as acid rain. Natural breakdown of uranium in the geoenvironment emits cancer causing radon gas into atmosphere.

IMPACT OF CONTAMINATION ON GEOENVIRONMENT

- In most of the cases, wastes are disposed off indiscriminately in low-lying areas without taking adequate engineering measures to effectively contain it.

- This results in a highly unhygienic and unhealthy environment leading to breeding of pests, mosquitoes and several harmful microorganisms.

- During precipitation, or groundwater coming in contact with these wastes generates contaminated water called leachate that can travel far field and pollute the surface and groundwater resources.

- Many of the harmful heavy metals can also travel along with the leachate if it is not contained properly.

- One of the complexities of contamination impact is its long term effects without a chance for realization. Most of the impacts are realized much later from rigorous studies, and by the time the damage would have been done.

- Hence, remediation becomes a tedious and cost-intensive affair. This makes geoenvironmental engineering a challenging and much needed subject. There is a need to focus on research that would help to predict and minimize the long term impact of indiscriminate and mismanaged waste contamination.

CASE HISTORIES ON GEOENVIRONMENTAL PROBLEMS

(i) Use of readily available local soil instead of expensive commercial soil (like bentonite) for waste Management

- Engineered waste management scheme necessitates the construction of highly impermeable barrier so that waste disposed on it does not find its way to ground water resources.

- Mostly these barriers are made of high plastic clays which are commercially available.

- This would considerably increase the cost of such geoenvironmental projects.

- Exploring the possibility of using local soils for such applications, therefore, becomes an important geoenvironmental problem. Any success in this direction would add to the economy of the project. This in turn would result in sustainable development of such very important project. e.g.

- Taha and Kabir (2005) have explored the possibility of using tropical residual soil for waste management, which is readily available over a considerable part of peninsular Malaysia.

- Hydraulic conductivity is used as the criterion for evaluation of soil suitability for the said application.

- The soil was compacted at different water content and compaction effort and then permeated with de-aired tap water. The results of hydraulic conductivity test indicates that the required flow of less than 10-9 cm/s can be achieved by using a broad range of water content and compaction effort.

- The soil has minimum shrinkage potential and adequate strength to support the load of waste overburden.

- These properties discussed would fall under the purview of geotechnical engineering. But the evaluation of soil suitability is not complete without understanding its chemical reactivity.

- In this study, cation exchange capacity (CEC) of soil is used as an indicator of chemical reactivity. It is desirable that the pollutants released from the waste disposal site should be effectively attenuated by the liners. This means that the soil should have high chemical reactivity. A soil with high CEC indicates high reactivity and hence high attenuation capacity of pollutants.

Knowledge of **soil-water interaction** and **soil-water-contaminant** interaction is very important for solving several problems encountered in geoenvironmental engineering projects.

Soil Mineralogy Characterization And Its Significance In Determining Soil Behaviour

- Soil is formed by the process of weathering of rocks which has great variability in its chemical composition.

- Therefore, it is expected that soil properties are also bound to the chemical variability of its constituents.

- Soil contains almost all type of elements, the most important being oxygen, silicon, hydrogen, aluminum, calcium, sodium, potassium, magnesium and carbon (99 percent of solid mass of soil).

- Atoms of these elements form different crystalline arrangement to yield the common minerals with which soil is made up of.

- Soil in general is made up of minerals (solids), liquid (water containing dissolved solids and gases), organic compounds (soluble and immiscible), and gases (air or other gases).

Formation Of Soil Minerals

- Based on their origin, minerals are classified into two classes: primary and secondary minerals.

- Primary minerals are those which are not altered chemically since the time of formation and deposition. This group includes quartz (SiO_2), feldspar ((Na,K)

AlSi3O8 alumino silicates containing varying amounts of sodium, potassium), micas (muscovite, chlorite), amphibole (horneblende: magnesium iron silicates) etc.

◉ Secondary minerals are formed by the decomposition and chemical alteration of primary minerals. Some of these minerals include kaolinite, smectite, vermiculite, gibbsite, calcite, gypsum etc.

◉ These secondary minerals are mostly layered alumino-silicates, which are made up of silicon/oxygen tetrahedral sheets and aluminum/oxygen octahedral sheets.

◉ Primary minerals are non-clay minerals with low surface area (silica minerals) and with low reactivity. These minerals mainly affect the physical transport of liquid and vapours.

◉ Secondary minerals are clay minerals with high surface area and high reactivity that affect the chemical transport of liquid and vapours.

◉ Silica minerals are classified as tectosilicates formed by SiO4 units in frame like structure.

◉ Quartz, which is one of the most abundant minerals comprises up to 95 percent of sand fraction and consists of silica minerals.

◉ The amount of silica mineral is dependent upon parent material and degree of weathering.

◉ Quartz is rounded or angular due to physical attrition. The dense packing of crystal structure and high activation energy required to alter Si-O-Si bond induce very high stability of quartz. Therefore, the uncertainty associated with these materials is minimal. In the subsurface, quartz is present in chemically precipitated forms associated with carbonates or carbonate-cemented sandstones.

◉ Clay minerals, which can be visualized as natural nano materials are of great importance to geotechnical and geoenvironmental engineers due to the more complex behaviour it exhibits. Therefore, emphases is given more on understanding clay mineral formation and its important characteristics. Basic units of clay minerals include silica tetrahedral unit and octahedral unit depicted in Fig. 4.

◉ It can be noted from the Fig. 4 that metallic positive ion is surrounded by non-metallic outer ions.

◉ Fig. 5 shows the formation of basic layer from basic units indicated in Fig. 4. There are 3 layers formed such as (a) silicate layer, (b) gibbsite layer and (c) brucite layer.

◉ Gibbsite layer is otherwise termed as dioctahedral structure in which two-third of central portion is occupied by Al^{+3}.

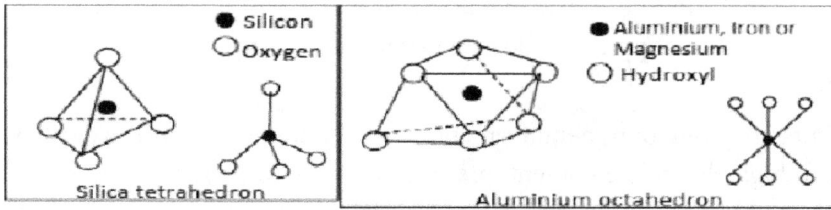

Fig. 4 Basic units of clay minerals

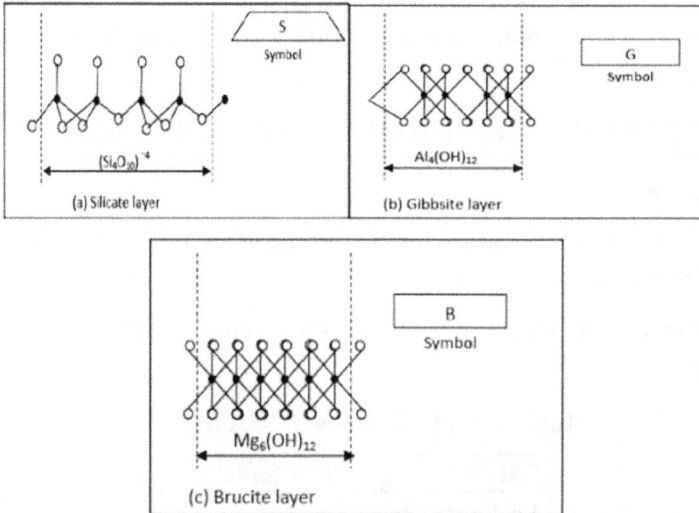

Fig. 5 Basic layer of mineral formation (modified from Mitchell and Soga 2005)

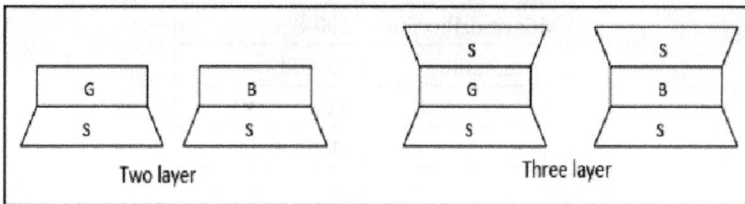

Fig. 6 Fundamental layers of clay minerals

⊙ Similarly, brucite layer is termed as trioctahedral structure in which entire central portion is occupied by Mg^{+2}.

⊙ These basic layers stack together to form basic clay mineral structure. Accordingly, there is two and three layer configuration as indicated in Fig. 6. More than hundreds of these fundamental layer join together to form a single clay mineral.

⊙ Kaolinite formation is favoured when there is abundance of alumina and silica is scarce.

⊙ The favourable condition for kaolinite formation is low electrolyte content, low pH and removal of ions that flocculate silica (such as Mg, Ca and Fe) by leaching. Therefore, there is higher probability of kaolinite formation in those regions with heavy rainfall that facilitate leaching of above cations.

- Similarly halloysite is formed due to the leaching of feldspar by H_2SO_4 produce by the oxidation of pyrite. Halloysite formations mostly occur in high-rain volcanic areas.

- Smectite group of mineral formation are favoured by high silica availability, high pH, high electrolyte content, presence of more Mg^{+2} and Ca^{+2} than Na^+ and K^+. The formation is supported by less rainfall and leaching, and where evaporation is high (such as in arid regions).

- For illite formation, potassium is essential in addition to the favourable conditions of smectite.

IMPORTANT PROPERTIES OF CLAY MINERALS

High Surface Area

- Specific surface area (SSA) is defined as the surface area of soil particles per unit mass (or volume) of dry soil. Its unit is in m^2/g or m^2/m^3.

- Clay minerals are characterized by high specific surface area (SSA) as listed in Table 1.

Table 1 Typical values of SSA for soils

Soil	SSA (m^2/g)
Kaolinite	10-30
Illite	50-100
Montmorillonite	200-800
Vermiculite	20-400
Silt	0.04-1
Sand	0.001-0.04

- High specific surface area is associated with high soil-water-contaminant interaction, which indicates high reactivity.

- For the purpose of comparison, SSA of silt and sand has also been added in the table. There is a broad range of SSA values of soils, the maximum being for montmorillonite and minimum for sand. As particle size increases SSA decreases.

- For smectite type minerals such as montmorillonite, the primary external surface area amounts to 50 to 120 m^2/g. SSA inclusive of both primary and secondary surface area (interlayer surface area exposed due to expanding lattice) and termed as total surface area would be close to 800 m^2/g.

- For kaolinite type minerals there is possibility of external surface area where in the interlayer surface area does not contribute much. There are different methods available for determination of external or total specific surface area of soils.

Plasticity And Cohesion

Clay attracts dipolar water towards its surface by adsorption. This induces plasticity in clay. Therefore, plasticity increases with SSA. Water in clays exhibits negative pressure due to which two particles are held close to each other. Due to this, apparent cohesion is developed in clays.

Surface Charge And Adsorption

Reasons for clay surface charge (i) Isomorphous substitution:

- During the formation of mineral, the normally found cation is replaced by another due to its abundant availability, e.g. when Al^{+3} replace Si^{+4} there is a shortage of one positive charge, which appears as negative charge on clay surface. Such substitution is therefore the major reason for net negative charge on clay particle surface.

- Dissociation of hydroxyl ions or broken bonds at the edges is also responsible for unsatisfied negative or positive charge.

Typical charged clay surface

- Positive charge can occur on the edges of kaolinite plates due to acceptance of H^+ in the acid pH range. It can be negatively charged under high pH environment.

- Absence of cations from the crystal lattice also contributes to charge formation.

- In general, clay particle surface are negatively charged and its edges are positively charged.

- Due to the surface charge, it would adsorb or attract cations (+ve charged) and dipolar molecules like water towards it. As a result, a layer of adsorbed water exists adjacent to clay surface.

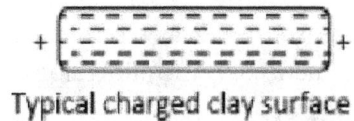

EXCHANGEABLE CATIONS AND CATION EXCHANGE CAPACITY

- Due to negative charge, clay surface attracts cations towards it to make the charge neutral. These cations can be replaced by easily available ions present in the pore solution, and are termed as exchangeable ions.

- The total quantity of exchangeable cations is termed as cation exchange capacity, expressed in milliequivalents per 100 g of dry clay.

EXCHANGEABLE CATIONS AND CATION EXCHANGE CAPACITY

- Due to negative charge, clay surface attracts cations towards it to make the charge neutral. These cations can be replaced by easily available ions present in the pore solution, and are termed as exchangeable ions.

- The total quantity of exchangeable cations is termed as cation exchange capacity, expressed in milliequivalents per 100 g of dry clay.

- Cation exchange capacity (CEC) is defined as the unbalanced negative charge existing on the clay surface.

- Kaolinite exhibits very low cation exchange capacity (CEC) as compared to montmorillonite.

- Determination of CEC is done after removing all excess soluble salts from the soil. The adsorbed cations are then replaced by a known cation species and the quantity of known cation required to saturate the exchange sites is determined analytically.

APPLICATIONS OF SOIL MINERAL ANALYSIS IN GEOENVIRONMENTAL ENGINEERING

The soil-water and soil-water-contaminant interaction and hence reactivity is greatly influenced by the mineralogy.

Soil-Water-Contaminant Interaction

- Under normal conditions, water molecules are strongly adsorbed on soil particle surface.

- Unbalanced force fields are generated at the interface of soil-water which increases soil-water interaction.

- When particles are finer, magnitude of these forces are larger than weight of these particles.

- This is mainly attributed to low weight and high surface area of fine particles.

Forces Between Soil Solids

- There are essentially two type of bonding: (1) Electrostatic or primary valence bond and (2) Secondary valence bond.

- Atoms bonding to atoms forming molecules are termed as primary valence bond. These are intra-molecular bonds.

- When atoms in one molecule bond to atoms in another molecule (intermolecular bond), secondary valence bonds are formed. What is more important in terms of soil solids is the secondary valence bonds.

- Van der Waals force and hydrogen bonds are the two important secondary valence forces.

- Secondary valence force existing between molecules is attributed to electrical moments in the individual molecules.

- When the centre of action of positive charge coincide with negative charge, there is no dipole or electric moment for the system and is termed as non-polar. However, for a neutral molecule there can be cases where the centre of action of positive and negative charge does not coincide, resulting in an electric or dipole moment. The system is then termed as polar.

- For example: water is dipole. Also, unsymmetrical distribution of electrons in silicate crystals makes it polar.

- Non-polar molecules can become polar when placed in an electric field due to slight displacement of electrons and nuclei. This is induced effect and the extent to which this effect occurs in molecule determines its polarisability.

- Van der Waals force is the force of attraction between all atoms and molecules of matter. This force comes into effect when the particles are sufficiently close to each other.

- Hydrogen bond is formed when a hydrogen atom is strongly attracted by two other atoms,

- For example: water molecules. This bond is stronger than van der Waals force of attraction and cannot be broken under stresses that are normally experienced in soil mechanics.

- These secondary valence bonds play a vital role in understanding soil-water interactions. Essentially, the forces dealt in soil mechanics may be grouped as gravitational forces and surface forces.

- From classical soil mechanics perspective, gravitational forces which are proportional to mass are more important.

- In geoenvironmental engineering surface forces are important. Surface forces are classified as attractive and repulsive forces. Attractive forces include (a) Van der Waals London forces (b) hydrogen bond (c) cation linkage (d) dipole cation linkage (e) water dipole linkage and (f) ionic bond.

- Van der Waals London force is the most important in soils and becomes active when soil particles are sufficiently close to each other.

- For example, fine soil particles adhere to each other when dry. Cation linkage acts between two negatively charged particles as in the case of illite mineral structure.

SOIL-WATER INTERACTION

- Water present in pore spaces of soil is termed as soil water or pore water. The quantity of water present in the pores will significantly influence its physical, chemical and engineering properties.

- It can be classified as (a) free water or gravitational water and (b) held water or environmental water.

⊙ As the name suggests, free water flows freely under gravity under some hydraulic gradient and are free from the surface forces exerted by the soil particle. This water can be removed by laboratory oven drying procedure.

⊙ Environmental water is held under the influence of surface forces such as electrochemical forces or other physical forces.

⊙ Both type of water are important in geoenvironmental engineering. There are many cases like seepage and infiltration problems whose solution necessitates the knowledge of free water.

⊙ However, these concepts are discussed in detail in classical soil mechanics text books.

⊙ The mechanism of soil-held water interaction is complex and influenced by soil type, mineralogy, current and past environmental conditions, stress history etc.

⊙ Held water can be further subdivided into structural water, adsorbed water and capillary water. Structural water is present within the crystal structure of mineral. This water is not very important as far engineering property of soil is concerned. For finding solution to several problems in geoenvironmental engineering, it is essential to understand in detail adsorbed water and capillary water.

Different Soil-Water-Contaminant Interaction Mechanisms

⊙ The contaminant that can pose serious threat to humans persist in short or long interval of time.

⊙ These contaminants can be naturally occurring ones such as arsenic, fluoride, traces of mercury or anthropogenic substances such as chlorinated organics, dissolved heavy metals etc.

⊙ The major role of a geoenvironmental engineer is to predict the fate of contaminants in the subsurface and minimize its migration towards groundwater.

⊙ Fate prediction is very essential to understand the presence of contaminants in groundwater sources or subsurface for long term (50 to 200 years).

⊙ This would essentially depend on different interaction mechanisms between contaminant and soil solids and also between contaminant and dissolved solutes present in pore water.

⊙ This knowledge is required to assess the risk or threat posed by these contaminants to humans and other organisms. Fate of contaminant in geoenvironment is decided by retention and transport of contaminants.

⊙ The important mechanisms governing these factors are as follows:

A. Chemical Mass Transfer And Attenuation

    a. Sorption- contaminant partitioning

    b. Dissolution/ precipitation- addition or removal of contaminants

    c. Acid-base reaction- proton transfer

    d. Redox reaction- electron transfer

    e. Hydrolysis/ substitution/ complexation/ speciation-ligand-cation complexes.

B. Mass Transport

    a. Advection- fluid flow

    b. Diffusion- molecular migration(c) Dispersion- mixing

C. Other Factors

    a. Biological transformations

    b. Radioactive decay

◉ An adequate knowledge of these mechanisms is required to predict the fate of contaminant.

◉ When the contaminated pore fluid passes through the soil mass, it is bound to undergo weak or strong reactions.

◉ Sorption process in which the contaminants clings on to the soil solids is one of the predominant reactions. Such reactions does not ensure permanent removal of contaminants from the pore fluid, rather attenuation takes place.

◉ Attenuation is the reduction in contaminant concentration during fluid transport due to retardation, retention and dilution. The extent of interaction between the contaminants and soil fraction determines reversible or irreversible nature of contaminant partitioning.

◉ The term retention is used for strong sorption of contaminants on the soil particles such that the concentration of pore fluid decreases with time.

◉ The amount of contaminant concentration reaching a particular target is considerably less than the source concentration. Chemical mass transfer and irreversible sorption removes the contaminants from the moving pore fluid.

◉ This is a very important aspect for a contaminant barrier system, where in the contaminants reaching ground water is minimized.

◉ Retardation is mainly governed by reversible sorption and hence release of contaminant would eventually occur. This will ensure the delivery of the entire contaminant load to the final target (example ground water), but with much delay.

◉ In nature, the effect of contaminated pore fluid is reduced when it interacts with fresh water (especially during precipitation). This process of dilution also delays the contaminant migration. However, the process of dilution is mostly independent of soil interaction.

◉ The process of retention and retardation is depicted in Fig. 7. From the figure, it can be noted that for retention process, the area under the curves (concentration) goes on reducing. For retardation, the area remains constant (mass conservation), however the concentration of a particular contaminant reduces.

For an effective waste management, retention process is more ideal than retardation. For proper prediction of contaminant fate, it is very essential to know whether the contaminant is retained or retarded.

Radioactive Decay

In this process, unstable isotopes decay to form new ones with release of heat and particles from element nucleus. The process is known as α or β decay depending on whether the element looses α particle (helium) or a β particle (electron). The process of decay is irreversible and daughter isotope increases in quantity. The disposal of radioactive waste generated from nuclear installations, mining etc. to subsurface will considerably increase the heat. Moreover, the radioactive isotope such as uranium, plutonium, cesium etc gets transported to far field and would pollute the groundwater. Preventing such harmful pollution and reducing the ill effect of overheating of subsurface is a challenging geoenvironmental problem.

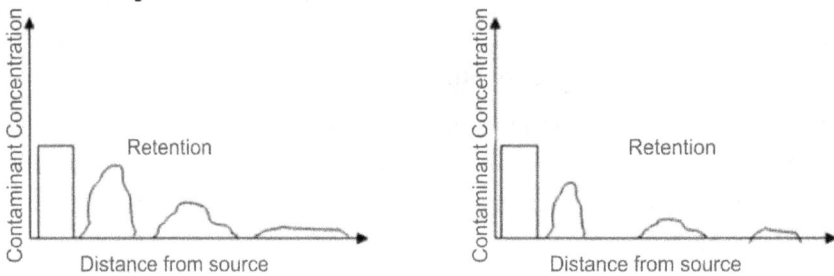

Fig. 7 Attenuation Process Due To Soil-Contaminant Interaction

Sorption And Partitioning

When contaminant laden pore water flow past the soil surface, mass transfer of these contaminants takes place on to the solids. The process is referred to as sorption or partitioning. The amount of partitioning depends on the soil surface (sorbent) and the reactivity of contaminant (sorbate). This is one of the predominant mechanisms governing the fate of contaminant once it is released into the geoenvironment. The term sorption refers to the adsorption of dissolved ions, molecules or compounds on to the soil surface. The mechanism of sorption includes physical and chemical sorption as well as precipitation reaction. These reactions are governed by surface properties of soil, chemistry of contaminant and pore water, redox potential and pH of the environment. Physical adsorption refers to the attraction of contaminant on to the soil surface mainly due to the surface charge (electrostatic force of attraction). Physical sorption is weak bonding and can be reversed easily by washing with extracting solution. Chemical sorption is strong force of attraction due to the formation of bonds such as covalent bond. High adsorption energy is associated with chemical sorption and it can be either exothermic or endothermic reaction.

Biological Transformation

Biological transformation is the degradation or assimilation of contaminants (mostly organic) by microorganisms present in the soil. Transformations from biotic processes occur under aerobic or anaerobic conditions. The transformation products obtained from each will be different. The biotic transformation processes under aerobic conditions are oxidation reaction. The various processes include hydroxylation, epoxidation, and substitution of OH groups on molecules. Anaerobic biotic transformation processes are mostly reduction reaction, which include hydrogenolysis, H^+ substitution for Cl^- on molecules, and dihaloelimination.

EVOLUTION OF WASTE CONTAINMENT AND DISPOSAL PRACTICES

- Increased events of environmental pollution and its realization have led to the evolution of planned and engineered waste management facilities.

- The waste management essentially comprises of collection, transport, disposal and/or incineration of wastes.

- A sustainable waste management is founded on 3 R's, namely Reduce, Reuse and Recycle so that the quantity of waste to be disposed on land is considerably reduced.

- The major focus is to reduce the quantity of waste production by efficient process control, try to reuse the by-products or waste products from a process, and try to recycle the left out waste products by value added transformation.

- Some of the major challenges faced in the implementation of an efficient waste management scheme are the non-awareness of public and the need for systematic functioning of various divisions like collection, transportation, disposal and site management.

- The concept of waste management started in 1800 century. However, the need for an integrated solid waste management program (ISWMP) has been realized in late 1980s.

- The main aim of ISWMP is to optimize all aspects of solid waste management to achieve maximum environmental benefits cost-effectively. It essentially consists of :-

1. Waste source identification and characterization.
2. Efficient waste collection
3. Reduction of volume and toxicity of waste to be discarded.
4. Land disposal and/or incineration.
5. Optimization of first four steps to reduce cost and environmental impact.

- The wastes which are produced include non-hazardous municipal solid waste, construction and demolition waste, partially hazardous medical wastes, agricultural waste, highly hazardous industrial and nuclear waste. The philosophy of handling hazardous and non-hazardous waste varies a lot.

- When the wastes are disposed on to the land, the percolating rainwater interacts with it and produces liquid known as leachate (contaminated liquid that comes out of the waste matrix).

- In the due course of time, the leachate percolates through the soil and reaches the groundwater and moves along with it as shown in Fig. 8.

- In the past, it was presumed that leachate generated from waste dumped directly on natural soil is completely attenuated (purified) by the subsurface before it reaches or interacts with ground water.

- In the figure, subsurface is the unsaturated natural soil which provides an indirect containment of harmful contaminants leaching out.

- In view of the above, all forms of non-engineered land disposal such as gravel pits were acceptable.

- Since, 1950 onwards there were considerable increase in the ground water pollution. The cause for such pollution was traced back to such indiscriminate casual waste disposals.

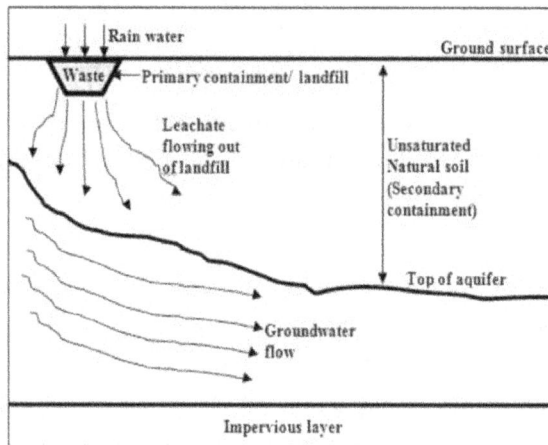

Fig. 8 A conceptual waste disposal facility on a global scale

- This gave way to the development of engineered waste disposal facilities known as landfills.

- The properties of soils used for the construction of landfills and the natural soil beneath the landfill become very important. So, major emphasis is laid on understanding the concepts of landfill and the role of soil in minimizing the harmful pollution of geoenvironment and ground water.

Landfills

- ◉ There are two types of landfills namely natural attenuation landfill and containment landfill.

- ◉ Natural attenuation landfill is similar to what has been discussed earlie where there is no provision below the wastes to minimize the migration of harmful contaminants. The unsaturated subsurface below the wastes naturally attenuate harmful contaminants before it reaches ground water. It is presumed that the contaminants reaching ground water will be well within the permissible limit, even though in most of the cases it would not be. For the same reason, these types of landfills are not preferred in spite of its simplicity.

- ◉ In the containment landfill, there is an engineered layer of soil known as liner on which the waste is disposed or dumped. Soil liners are constructed with some desirable properties meeting the regulations set by the pollution control board.

- ◉ The design of these liners is done in such a way that the contaminants leaching out seeps at a very low pace and gets attenuated.

- ◉ The concentration of contaminants reaching the ground water within the prescribed design life is expected to be well within the permissible limit. This type of landfill is mandatory for containing hazardous wastes such as industrial and nuclear wastes. All the working elements of such landfills are properly designed.

Engineered Landfills

- ◉ The first and foremost task in the planning of engineered landfills is its site selection.

- ◉ There are several socio-economic concerns which need to be satisfied before a site can be decided for waste disposal.

- ◉ The major concern is social since nobody likes wastes to be dumped in their neighborhood. This would necessitate mass education and awareness program on the pros and cons of the waste management project.

- ◉ Apart from public acceptance the other factors which are important in site selection are locational, geotechnical and hydrogeological criteria.

- ◉ Another important aspect in landfill site selection is establishing search radius, which is the maximum distance of waste hauling (transport). Waste hauling is one of the costliest items in landfill operations.

- ◉ Three important steps of landfill site selection are
    1. Data collection
    2. Locational criterion
    3. Obtaining public reaction and acceptance

a. Data collection: The data pertained to landfill site selection are summarized as follows:

i. Topographic maps: This include information on contour, natural surface, water drainage, location of streams, wetlands etc. Ideally landfills should be avoided on land contributing to groundwater recharge. The surface flow should be in such a way that water flow away from the landfill site. In case the flow is towards the landfill then adequate measure has to be taken to prevent excessive water seeping into the landfill.

ii. Soil Maps: Gives information on the type of soil available at a particular place. This information is important before going for an in depth subsurface investigation. A high permeable soil strata is normally avoided for landfills.

iii. Land Use Maps: These maps are very important as it gives the land value and its importance. There will be some zoning restriction for some lands laid down by the government, which can be assessed based on land use maps. For example, landfills should be located away from the flood plain.

iv. Transportation: The data on transportation would include the present network and the futuristic development. It is very essential that the landfill site is easily accessible and waste hauling is optimal. At the same time, the site should be away from important facilities like airport. It is essential to consider road and rail network details before site selection.

v. Waste Type And Volume: The primary question is whether the waste is hazardous or not.

⊙ The philosophy of waste containment changes depending on whether it is municipal or industrial waste. Stringent specifications need to be followed for industrial waste and in no case the waste can be dumped in open pits.

⊙ Around 50% of the total waste comes from domestic municipal sources. A waste generation rate of 0.9-1.8 kg/person/day is a reasonable estimate for determining municipal waste volume. The population and its growth during the active life of landfill need to be computed.

⊙ Waste volume per year = population per year x waste generation rate.

⊙ The landfill volume is the sum of daily, intermittent and final cover volume and waste volume. Waste: daily cover ratio of 4:1 is needed if soil is used as the cover.

B. Locational Criterion: Following are some of the important points to be followed while deciding location for waste containment.

- **Lake Or Pond:** Away by 300m. The distance can be reduced for engineered waste containment. Surface water need to be monitored continuously for pollution in future.

- **River:** Away by 100 m.

- **Highway and public park:** Away by 300 m.

- **Airport:** Away by 3 km to avoid bird menace.
- **Water supply well:** Away by 400 m.
- Crowded habitat, wetland, unstable area to be avoided.
- The geology of the place should be suitable with no faults and folds. Maximum horizontal acceleration for the site caused by earthquake should not exceed 0.1g in 250 years.

C. Preliminary Assessment Of Public Reaction:

- Public education on the short term and long term advantages of the facility should be carried out extensively.
- Not in my backyard (NIMBY) sentiment can prevent the execution of landfill.
- Some of the major concerns are noise, odour, increase in traffic volume, reduction in property value, fear of groundwater contamination etc.
- The public needs to be assured that the above mentioned concerns would be tackled efficiently. This is one of the challenging issues for geoenvironmental engineers and municipal authorities in the planning and execution of such projects.

Engineered Containment Landfills

- The engineered landfill includes designed man made barrier layers for minimizing the migration of harmful contaminants from the place of disposal to the groundwater.

Major Role Of Soil In Engineered Landfill

- As indicated in Fig. 9, the major role of soil in an engineered landfill can be summarized as follows:

1. Compacted liner or barrier which minimize the migration of contaminant to groundwater and hence it is the most integral and important part of a landfill. The reduction in migration is due to low permeability and contaminant retention capacity of the clayey soil used in liners.

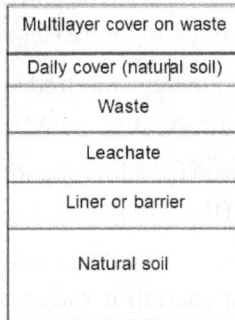

Multilayer cover on waste
Daily cover (natural soil)
Waste
Leachate
Liner or barrier
Natural soil

Fig. 9 Role of soil in an engineered landfill

2. Leachate collection system provided below the waste to collect the leachate and effectively drain to a collection source for further treatment.

3. After the service life of the landfill, an integrated multi layer cover system is provided on top of the waste to isolate it from the environment and minimize the generation of post closure leachate.

4. Natural soil is used as daily cover material for waste during the operational phase of landfill.

5. The unsaturated natural soil below the liner act as an additional buffer layer in reducing the migration of contaminants to groundwater.

6. In addition, suitable geosythetics, geotextiles, geomembrane, geonets etc. are used individually or in combination with soil to act as liner, drainage layer, filtration layer or separation layer. The use of geosynthetic helps to reduce the thickness of liner layer.

7. Compacted liner or barrier which minimize the migration of contaminant to groundwater and hence it is the most integral and important part of a landfill. The reduction in migration is due to low permeability and contaminant retention capacity of the clayey soil used in liners.

8. Leachate collection system provided below the waste to collect the leachate and effectively drain to a collection source for further treatment.

9. After the service life of the landfill, an integrated multi layer cover system is provided on top of the waste to isolate it from the environment and minimize the generation of post closure leachate.

10. Natural soil is used as daily cover material for waste during the operational phase of landfill.

11. The unsaturated natural soil below the liner act as an additional buffer layer in reducing the migration of contaminants to groundwater.

12. In addition, suitable geosythetics, geotextiles, geomembrane, geonets etc. are used individually or in combination with soil to act as liner, drainage layer, filtration layer or separation layer. The use of geosynthetic helps toreduce the thickness of liner layer.

Subsurface Investigation For Waste Management

- Subsurface investigation for waste management is required for deciding the site for landfills and also for delineating the extent of contamination.

- Several hydrogeological parameters required for landfill site selection are obtained from subsurface investigation conducted for different potential sites.

- The methodology for subsurface investigation remains similar to any other geotechnical investigation (for example, open pit, bore holes).

- In addition, several geophysical methods such as electrical resistivity imaging, seismic refraction, ground penetration radar, etc. are used for defining the zone of contamination, establishing the depth of aquifer, and also to reduce the number of bore holes.

Design of Landfills

- An engineered landfill essentially consists of a barrier layer or liner which is a low permeable zone to prevent the leaching of waste from the landfill.

- Above the liner, a drainage layer is placed which collects the leachate from the waste for treatment. Such a layer also minimizes the head causing flow in liner due to the timely removal of leachate from the landfill.

- The third important layer is the cover to the landfill, which is a multi-layered system to cut off the harmful effect of waste on the atmosphere. The various aspects required for planning and design of landfill are as follows:

  1. Waste Characterization
  2. Assessment of leachate and gas generation
  3. Landfill elements to be provided
  4. Liner and cover materials
  5. Landfill design approach

Waste Characterization

Waste characterization is important to understand the following:

1. Physical and chemical tests are preformed to evaluate whether waste is hazardous or non-hazardous.
2. Whether waste can be landfill directly or necessitate processing (reduction, recycling etc.) before disposal.
3. Approximate rate of waste volume generated.
4. Assessment of leachate quantity.
5. Assessment of leachate quality for judging liner compatibility, treatment plant design, ground water monitoring program design.
6. Safety precautions to be followed during landfill operations.
7. Identify waste reduction alternatives.

Assessment Of Leachate And Gas Generation

- Leachates are produced when water or other liquids percolates and interacts with waste.

- The information on quality and quantity of leachate and gas generated during active life and after closure are important for realistic and efficient design of a landfill. Leachate contains a lot of dissolved and suspended materials.

- Gases produced include CH_4, CO_2, NH_3 and H_2S due to anaerobic decomposition of waste. These gases either escape to atmosphere or dissolve in water leading to further reactions.

- Contaminated liquids of high concentration are formed due to chemical reaction taking place within the waste. The percolating water increases the quantity of leachate but would help to dilute the concentration.

FACTORS INFLUENCING LEACHATE QUALITY

- Refuse composition

- Elapsed time: Leachate quality (concentration) increases and reaches peak during the working period of landfill and then start decreasing with time. All the contaminants present in the leachate do not exhibit peak at the same time and may not be of same shape.

- Temperature: Temperature affects bacterial growth and chemical reactions, there by affecting leachate quality.

- Available moisture influences biodegradable and subsequent leaching of wastes. rate is required to plan the management of leachate and cost incurred for it.

  Leachate quantity considerably reduces after closure and construction of covers. Leachate volume (Lv) is given by Eq. 1.

  $$L v = P + S - E - AW \qquad 1$$

  Where P is the precipitation volume, S is the volume of pore liquid squeezed from the waste, E is the volume lost by evaporation and AW is the volume of liquid lost through absorption in waste. Pore Squeeze Leachate Volume (S)

- When sludge is disposed, liquid within the pores gets squeezed due to self- weight of sludge and weight of waste dump and cover soil.

- Such an action is similar to the consolidation process occurring in a saturated soil. Primary consolidation of waste accounts for the majority of pore squeeze leachate. The primary consolidation properties of sludge are used to predict leachate generation rate.

- Loss due to evaporation depends on ambient temperature, wind velocity difference in vapour pressure etc.

- Leachate absorbed in waste (AW) is depended on field capacity (FC) of waste. FC is the maximum moisture content that waste can retain against gravitational force without producing down ward flow. When the moisture content is within FC, the waste has the capacity to retain water without causing downward flow.

Post Closure Leachate Generation Rate

- Only water that can infiltrate through the final cover of the landfill percolates through the waste and generates post closure leachate.

- Water balance method expressed by Eq. 2 is a popular method for estimating post closure leachate generation.

  $$L'V = P - ET - R - S \qquad 2$$

  Where L'V is the volume of post closure leachate, P is the volume of precipitation, ET is the volume lost though evapotranspiration, R is the volume of runoff and S is the volume of moisture stored in soil and waste. Potential ET is obtained based on appropriate empirical equation.

R = Cr I A 3

Where Cr is the run off coefficient, I is the rainfall intensity and A is the area of landfill surface.

◉ Soil moisture storage (S): A portion of infiltrating water is stored by soil and only a part of this is used for vegetation. Soil moisture storage capacity is the difference between field capacity and wilting point.

◉ Wilting point is the moisture content at which plants cannot draw moisture and starts wilting. Normally, moisture content corresponding to 1500 kPa matric suction is taken as wilting point.

◉ Water balance method if not done properly results in large errors especially when used for long term leachate generations rate. The disadvantages of water balance method are:

   i. It does not account permeability of cover layer
   ii. Evapotranspiration is sometimes wrongly calculated due to over prediction of root length in vegetation layer. In reality root would not have penetrated entire thickness of vegetation layer.

Gas Generation Rate

◉ Gas generation rate is mostly valid for municipal solid waste (MSW) landfill where organic matter decomposition results in the production of gases.

◉ Gas production in MSW landfill occurs due to anaerobic degradation resulting from hydrolysis and fermentation (attributed to bacterial activities), acetogenesis and dehydrogenation, and methanogenesis.

◉ Hydrogen gas is produced due to the oxidation of soluble products to organic acids.

◉ Some of the other gases produced from MSW are methane, carbon dioxide, hydrogen sulphide and nitrogen.

◉ Gas production reaches a stable rate and then decreases as biological activity in MSW landfill start decreasing.

◉ The assessment of time dependent percentage production of methane from a MSW landfill is important for recovering methane as an energy source, and thereby reducing greenhouse gas effect.

Engineered Containment Landfills

◉ The engineered landfill includes designed man made barrier layers for minimizing the migration of harmful contaminants from the place of disposal to the groundwater.

⊙ The provisions in engineered landfill depends upon the type of waste is receives.

MAJOR ROLE OF SOIL IN ENGINEERED LANDFILL

1. Compacted liner or barrier which minimize the migration of contaminant to groundwater and hence it is the most integral and important part of a landfill. The reduction in migration is due to low permeability and contaminant retention capacity of the clayey soil used in liners.

2. Leachate collection system provided below the waste to collect the leachate and effectively drain to a collection source for further treatment.

3. After the service life of the landfill, an integrated multi layer cover system is provided on top of the waste to isolate it from the environment and minimize the generation of post closure leachate.

4. Natural soil is used as daily cover material for waste during the operational phase of landfill.

5. The unsaturated natural soil below the liner act as an additional buffer layer in reducing the migration of contaminants to groundwater.

6. In addition, suitable geosythetics, geotextiles, geomembrane, geonets etc. are used individually or in combination with soil to act as liner, drainage layer, filtration layer or separation layer. The use of geosynthetic helps to reduce the thickness of liner layer.

APPLICATION OF GEOSPATIAL TECHNOLOGY IN GEOENVIRONMENTAL ENGINEERING

Introduction

The soil contamination can be considered as the presence of man-made chemicals or other alteration in the natural soil environment. Some of the soil properties can be estimated using spectral responses of surface soil for different bands of radiometer. The great demand for information about properties of surface and sub-surface soil and short time available for the work arises to employ remote sensing techniques in geotechnical engineering which reduce the time, cost, efforts, and staff.

Geospatial technologies is a term used to describe the range of modern tools contributing to the geographic mapping and analysis of the Earth and human societies. These technologies have been evolving in some form since the first maps were drawn in prehistoric times. In the 19th century, the long important schools of cartography and mapmaking were joined by aerial photography as early cameras were sent aloft on balloons and pigeons, and then on airplanes during the 20th century. The science and art of photographic interpretation and map making was accelerated during the Second World War and during the Cold War it took on new dimensions with the advent of satellites and computers. Satellites allowed images of the Earth's surface and human activities therein with certain limitations.

Computers allowed storage and transfer of imagery together with the development of associated digital software, maps, and data sets on socioeconomic and environmental phenomena, collectively called geographic information systems (GIS). An important aspect of a GIS is its ability to assemble the range of geospatial data into a layered set of maps which allow complex themes to be analyzed and then communicated to wider audiences. This 'layering' is enabled by the fact that all such data includes information on its precise location on the surface of the Earth, hence the term 'geospatial'.

Especially in the last decade, these technologies have evolved into a network of national security, scientific, and commercially operated satellites complemented by powerful desktop GIS. In addition, aerial remote sensing platforms, including unmanned aerial vehicles (e.g. the GlobalHawk reconnaissance drone), are seeing increased non-military use as well.

High quality hardware and data is now available to new audiences such as universities, corporations, and non-governmental organizations. The fields and sectors deploying these technologies are currently growing at a rapid pace, informing decision makers on topics such as industrial engineering, biodiversity conservation, forest fire suppression, agricultural monitoring, humanitarian relief, and much more.

Geospatial technology is a significant scientific finding, which moved the possibilities of humankind to a brand-new level.

Using geospatial technology is comparatively inexpensive and simple, while its possibilities are next to unlimited. Applications of geospatial technologies are incorporated in almost any sector, industry, or research where the location is important.

Types of Geospatial Technologies

Geospatial technology **correlates an object's position with its geographic coordinates**. The idea is not new and served for observing places with pigeons or balloons first, primarily for mapmaking purposes. However, it is dramatically deployed in the era of satellites and computers

Identification of geospatial data enables monitoring, tracing, measuring, assessment, identification, or modeling. The basic list of geospatial technologies encompasses remote sensing (RS), GPS, and GIS.

Remote Sensing

Different types of remote sensing as geospatial technology enables us to study objects or surfaces at faraway distances employing their reflectance properties. Sensing them with active or passive systems, measuring and analyzing the response, experts can assess the target's properties and make corresponding conclusions.

Satellites revolve our planet and generate imagery based on several source options and methods of geospatial technology for data collection:

- ◉ **Electromagnetic impulses** (including visible, infrared, and microwave channels);
- ◉ **Filmed or digital areal imagery** from piloted and non-piloted vehicles (e.g., airplanes and drones);
- ◉ **Radars and lidars enabling** to calculate the distance with radio or light signals correspondingly.

Advanced systems distinguish objects of one meter and even smaller.

Global Positioning Systems (GPS)

GPS bases on the geometric phenomenon of triangulation. As the name suggests, calculations ground on three sources. It is a typical situation, however. When it relates to space and signals, scientists have to bear in mind that transmitted energy travels at the speed of light, causing possible calculation discrepancies. To minimize errors and to make the calculations more accurate, global positioning systems use four sources.

Geographic Information Systems (GIS)

GIS, one type of geospatial technology, merges spatial and non-spatial data, remote sensing imagery, GPS data points to elaborate a single complete system. It allows users to collect, group, and analyze required information on multiple layers, including elevation, vegetation species, forest health, roads, water bodies, animals, etc.

Why Is Geospatial Technology Important?

The innovation helps to find answers to many questions arising in multiple industries and sectors. At the dawn of its development, the data access and its application scale were limited. Nowadays, geospatial technology importance went far beyond cartographic or military needs.

Geospatial technology allows tracking a questioned object and referring it to a specific location. This feature helps people to complete scientific or non-scientific tasks, governmental and non-governmental, military and civil.

The importance of geospatial technology is equally recognized by common people and giant corporations. It serves to fulfill both strategic and minor tasks like tracking atomic submarines or sharing one's location with a friend.

APPLICATIONS OF GEOSPATIAL TECHNOLOGY

The scope of geospatial data use is vast: it embraces every sphere or industry where geographical position matters. The list includes geography proper, ecology, tourism, marine sciences, agriculture, forestry, marketing and advertising, military forces, navy, aircraft, law enforcement, logistics and transportation, astronomy, demography, healthcare, meteorology, and many others.

Here are some typical examples of how geospatial technology is applied:

- ◉ **Logistics.** Tracking goods and ensuring their quality.

- ◉ **Transportations.** Identifying location and time of arrival, route making, and navigation.

- ◉ **Meteorology.** Referring weather forecasts to particular territories.

- ◉ **Forestry.** Detecting forest fires and deforestation & preventing large-scale wildfires (read more about satellite monitoring of forest fires and deforestation in Brazil Amazon).

- ◉ **Agriculture.** Assessing vegetation state on a selected terrain.

- ◉ **Healthcare.** Monitoring areas of epidemic outbreaks.

- ◉ **Ecology.** Tracing species populations in certain areas, preventing and addressing calamities.

- ◉ **Marketing and advertising.** Targeting ads to relevant regions.

- ◉ **Real estate.** Visualizing and analyzing real estate objects remotely.

- ◉ **Insurance**. Managing risks in questioned areas (e.g., via historical georeferenced data analysis).

FUTURE OF GEOSPATIAL TECHNOLOGY

Even though it is difficult to imagine a sphere that does not use geospatial technologies, the finding prospects are even more promising. It assists in making weighted decisions and allows even more accurate analysis.

The technologies find new implementations, and related researches go further. They are affordable for a wide audience, and their practical use inspires a greater spectrum of applications in the future. The reason for their popularity is in data accuracy, which means better precision and, thus, increased productivity.

Geospatial technologies enhance the performance of artificial intelligence and smart machinery in multiple spheres and agriculture in particular. Remotely controlled equipment completes numerous tasks via GPS and digital dashboards. Robots and smart machinery in the fields seem futuristic no longer, and it is not the limit.

Expansion and new application solutions are expected in biosecurity, education, construction, engineering, ecology, food supplies, precision agriculture, financial market, statistics, transportation, to mention a few.

Basically, geospatial data enhances performance in each sphere, outlining specific needs or issues in selected regions. When it comes to farming, for example, landowners can save costs and efforts by treating only critical spots with exact coordinates on the field map and see a big picture of their farmlands at the same time.

Further employment of GPS in the automobile and aircraft industries enables frequent use of driverless vehicles and UAVs as a matter of fact.

The development of geospatial technologies brings quite an interesting correlation onto the scene. New achievements in this branch mean the corresponding upgrade of related industries. So, the improvement process is not likely to stop, ensuring even greater precision, credibility, performance, quality, and security.

Remote Sensing

All methods of collecting information about Earth without touching it are forms of remote sensing. For example, photographing a flooded river from an airplane is a form of remote sensing, but sampling its water is not. Scientists mount remote-sensing devices such as cameras and radars on ships, airplanes, or satellites that can cruise above the land, sea, or ocean floor, collecting data from large areas quickly. Most remote-sensing technologies use electromagnetic waves (light or radio waves) because they are fast, interact in revealing ways with solid matter, and can pass easily through both air and vacuum.

Active or Passive.

A given remote-sensing technology may be either active or passive. Passive remote-sensing technologies collect whatever waves happen to be coming their way and form them into an image. They usually are tuned to a particular band of the electromagnetic spectrum, such as visible light or infrared light (heat radiation). Active remote-sensing systems, on the other hand, beam radio pulses, sound waves, or laser beams at their targets and construct an image from the echoes.

Remote sensing is important for water science because water is widely distributed. Precipitation falls unevenly and intermittently over large areas; rivers drain irregularly-shaped watersheds; lakes are dotted randomly over vast territories. On-site measurements provide accurate information at single points, but aircraft or satellites observe entire landscapes at once. Furthermore, remote-sensing observations are easy to repeat; a satellite, for example, may pass over the same part of the Earth repeatedly. It is straightforward to image the same part of Earth repeatedly from space, making it possible to monitor both local and large-scale changes.

Remote-sensing imagery acquired by passive and active systems has been used for almost every conceivable type of water-science research, including the study of waves, currents, and oceanic circulation patterns; the detection of marine organisms; the inventorying of lakes; and the mapping of wetlands , ocean floors, land-water boundaries, floodplains , ice movement, snow cover, oil spills, shoreline changes, and many other features. Remote-sensed data are key in flood prediction and prevention, irrigation and drainage studies, water quality management, groundwater protection, and other water resource investigations.

Satellites

The Landsats (for *land* -sensing *sat* ellites) have for decades been the most important remote-sensing satellites, though many others also have been launched. The first Landsat was lofted by the United States in 1972 and the seventh was functioning as of 2002. Landsats circle Earth at a low altitude of 420 to 912 kilometers (260 to 570 miles), passing over the North and South Poles rather than circling the equator. This path means that Earth rotates inside the circular orbit of each satellite, constantly presenting new territory to the satellite's view. The data collected by the Landsats are made publicly available by the U.S. government.

Satellite images often record visible light or other forms of radiation. Visible-light images are useful for determining the locations and sizes of rivers, lakes, ice-covered or snow-covered areas, and other surface features. Images made from infrared light (which can be felt on the skin as radiant heat but not seen) provide different information. For example, infrared images can reveal heat plumes in bodies of water such as those occurring when a nuclear power plant disposes of hot water from its cooling system into a river. Creating detailed maps of such heat plumes by taking on-the-spot temperature readings would require thousands of measurements over a large area of moving water, an effectively impossible task. Infrared images also are used to classify vegetation cover by type, to monitor droughts, to visualize changes in ocean currents, and for other water-related purposes.

Radar

Radar works by sending radio pulses toward objects and recording how long the echoes take to return, and is the most common active remote-sensing technology. Radar is a superb tool for measuring relief (i.e., the height variations of land or water surfaces). Precise knowledge of topography (land relief) is essential to understanding the collection of rainwater and snowmelt by rivers, which is in turn useful for flood prediction, agriculture, and other purposes. Radar observations of water waves give information on winds and currents; satellite-based radar also has mapped subtle variations in sea level due to irregularities in Earth's gravitational field.

Radio waves pass readily through clouds and ice. Radar's ability to see through ice makes it an ideal tool for measuring the thickness of ice and snowfields. The Greenland and Antarctic ice caps, which contain nearly 87 percent of the world's fresh water, have been sounded with radar, leading to the discovery of over 70 Antarctic lakes. One, called Lake Vostok, is the size of Lake Ontario and is buried beneath kilometers of ice.

Radar studies of water and ice are not limited to Earth. The U.S. National Aeronautics and Space Administration (NASA) has proposed that spacecraft-mounted radar be used to peer through the icy crust that covers Europa, one of the moons of Jupiter, to see whether a global ocean lies hidden beneath it. The European Space Agency's Mars Express mission, scheduled for launch in 2003, was expected to carry radar to map ice deposits in the soils of Mars.

Sonar

Sonar, another active remote-sensing technology, is similar to radar but uses sound waves rather than radio waves. Most of the world's water, the 97 percent that is contained in the oceans, can be explored only superficially using light or radar because water absorbs electromagnetic waves. This makes sonar the only remote-sensing technology that can explore the bulk of the world's seas and other deep waters. Sonar systems mounted on ships' hulls or in special torpedoes towed behind ships ("sonar fish") have been used since World War II to map the ocean-floor topography of the world, and the knowledge thus gained has revolutionized the knowledge of geology.

Global Positioning System (GPS)

GPS is a system of twenty-four satellites that allows the coordinates of any point on or near Earth's surface to be measured with extremely high precision. The satellites of the GPS are arranged evenly around the Earth so that at least four are visible at all times from any point on Earth's surface. As one satellite sets below the horizon, another always rises somewhere else.

A special GPS receiver unit, usually handheld, is used to receive signals broadcast by the GPS satellites and to compute its own location from those signals. The GPS is a high-tech shortcut to the goal that surveyors and navigators have long sought by slower, less accurate means: namely, precise knowledge of one's own location. The GPS has only been in widespread use since more affordable GPS receivers first became available in the early 1990s.

The coordinates supplied by a GPS receiver must be matched to other data to be meaningful. A GPS receiver relays where something is located, but a human operator must specify what that something is: a rainfall measurement, well, or some other datum or object. The GPS is not a remote-sensing system because it does not collect data about the Earth. Rather, it facilitates the collection of data at specific points on Earth's surface.

GEOGRAPHIC INFORMATION SYSTEMS (GIS)

GIS is a computer system that can be used for scientific research and for water planning and management. This technology integrates powerful computer capabilities with the unique visual perspective of a good old-fashioned map.

It allows users to assemble, store, manipulate, and display geographically referenced information—data that can be identified according to its locations.

GIS requires the use of computer hardware, software, data, and specialists to study data related to Earth and the interconnections between its various features. GIS links information about where things are with information about what things are like. It is much more sophisticated than a paper map, as it can combine many layers of information about a particular spatial area to help yield a better understanding of that place. The layers of information to be studied depend on the researcher's purposes.

The information organized in a GIS comes from many sources, including remote sensing, the GPS, censuses, rock and soil samples, stream gage systems, weather stations, and wildlife sightings. GIS stores information about a geographic area as a collection of themed layers or maps. An individual layer can be anything that contains similar features (e.g., lakes, streets, postal codes, agricultural lands). Reliable and accurate data are necessary for creating high-quality GIS maps.

GIS is utilized in many industries, including business, transportation, communications, and defense; similarly, GIS is an important tool for analyzing water policy and science issues. ˙ Because all the layered maps in a given GIS cover the same geographic area, it is possible to link and compare different kinds of data. For example, information about a river network or watershed can be compiled using GIS. A watershed represents the landscape view of water resources, and the effects of spatial patterns of soils, land use, land cover, and urban development can be studied.

Combining topography with vegetation and soil type, for example, may help hydrologists understand how water drains from a given landscape, and thus enable them to predict flooding or pollution transport. GIS makes it possible for earth scientists, city planners, farmers, the military, water resource managers, and many other users to handle the vast quantity of information available from remote sensing, GPS, geology, biology, and other sources.

GIS and other geospatial technologies provide significant input for the management and analysis of large volumes of data, allowing for better understanding of environmental processes and for better management of human activities to ensure environmental quality and economic vitality.

9

WEATHERING - SOIL FORMATION FACTORS, PROCESSES AND COMPONENTS

Soil Science: "The science dealing with soil as a natural resource on the surface of the earth, including Pedology (soil genesis, classification and mapping), physical, chemical, biological and fertility properties of soil and these properties in relation to their management for crop production.

DEFINITIONS OF SOIL

Generally soil refers to the loose surface of the earth as identified from the original rocks and minerals from which it is derived through weathering process.

Whitney (1892): Soil is a nutrient bin which supplies all the nutrients required for plant growth

Hilgard (1892): Soil is more or less a loose and friable material in which plants, by means of their roots, find a foothold for nourishment as well as for other conditions of growth"

Dokuchaiev (1900): Russian scientist - Father of soil science - Soil is a natural body composed of mineral and organic constituents, having a definite genesis and a distinct nature of its own.

Joffe (1936): "Soil is a natural body of mineral and organic constituents differentiated into horizons - usually unconsolidated - of variable depth which differs among themselves as well as from the underlying parent material in morphology, physical makeup, chemical properties and composition and biological characteristics".

Jenny (1941): Soil is a naturally occurring body that has been formed due to combined influence of climate and living organisms acting on parent material as conditioned by relief over a period of time.

Soil Science has six well defined and developed disciplines

Soil fertility : Nutrient supplying properties of soil

Soil chemistry : Chemical constituents, chemical properties and the chemical reactions

Soil physics : Involves the study of physical properties

Soil microbiology : Deals with micro organisms, its population, classification, its role in transformations

Soil conservation: Dealing with protection of soil against physical loss by erosion or against chemical deterioration i.e excessive loss of nutrients either natural or artificial means.

Soil Pedology : Dealing with the genesis, survey and classification

Soil formation and development is a dynamic rather than static process. Soils were present when prehistoric animals roamed the Earth and, like those animals, some are no longer present or are preserved only as fossilized soils buried deep beneath our present soil.

Weathering describes the means by which soil, rocks and minerals are changed by physical and chemical processes into other soil components.

Weathering is an integral part of soil development. Depending on the soil-forming factors in an area, weathering may proceed rapidly over a decade or slowly over millions of years.

The development of a soil reflects the weathering process associated with the dynamic environment in which it has formed. Five soil-forming factors have been identified that influence the development of a specific soil. Wherever these five factors have been the same on the landscape, the soil will be the same. However, if one or more of the factors differ, the soils will be different. The factors are:

1. Parent material
2. Climate
3. Living organisms
4. Topography
5. Time

Parent Material

Parent material is made of rock and minerals. When the other four soil-forming factors act on parent material, it is weathered into smaller particles forming soil.

There are many types of parent material with different mineral contents. The Earth is believed to be about three billion years old. Mountains have been created and eroded away and then created again. Seas have covered the land and receded leaving layers of mud, sand and lime carbonate thousands of feet thick. Volcanoes have erupted. Glaciers have formed during long periods of cold weather and melted during long periods of warm weather.

Parent material can be rock formed in place or the remnants of rock that was moved by wind, water, ice or even gravity. A variety of parent material can be found in Nebraska ranging from sand in the Sandhill Region to clays in the Missouri and other river bottoms.

In the Great Plains, especially in the south, parent materials are primarily associated with ancient seas. These seas came into the region and receded several times, leaving sediment behind, which, over time, became sandstone, limestone and shale bedrock formations. Bedrock soil formations are classified as residuum parent materials and can be exposed and broken down to form soil.

Most residuum (e.g., limestone, sandstone and shale bedrock formations) is covered with more recent geological materials such as glacial deposits, windblown minerals or materials moved by water. One or more overlying parent materials may have been deposited in an area throughout time.

Glaciers are believed to have invaded only the eastern portion of Nebraska where they filled the valleys and leveled the hills. Sand and gravel were deposited along with boulders, clay and other sediments as the glaciers melted and retreated. Rivers which had previously flowed full of water were blocked and large amounts of sand and gravel washed from the Rocky Mountains were deposited in central Nebraska. These accumulations of sand and gravel are now the aquifers that provide our abundant supply of groundwater.

Many soils in southeastern Nebraska were formed in parent materials deposited by the glaciers, usually referred to as glacial drift, glacial till or glacial outwash.

Much of the parent material deposited in ancient times has been covered by windblown material. The windblown silty material is called loess. It covers most of Nebraska to varying depths, except in the Sandhills and western portions of the Panhandle. The yellow-brown loess is primarily found in the subsoil zone and may be 700 feet or more deep in the northeast and central areas of the state and only a few feet deep in western and southeast Nebraska. Loess soils are generally very fertile. Some are among the most productive soils in the world.

Windblown sand material is called eolian sand. It predominantly covers residuum in the Sandhills and western portions of the Panhandle. This coarse textured parent material is usually several feet deep and is found in both the surface and subsoil zones. Eolian soils are not very productive because they have very low water-holding capacity, are low in organic matter, and are nutrient deficient as compared to loess soils. Most are used for grass production or natural habitat.

Geologic materials moved from the parent material by water are known as alluvium. Alluvial deposits are found in flood plain areas such as the Platte River and other stream valleys. Since stream beds constantly change over time, alluvial parent materials are highly variable as are the soils that form them.

The physical and chemical weathering processes that change parent material into soil include:

- ⊙ Temperature changes — freezing and thawing.
- ⊙ Erosion by water, wind, ice and gravity.
- ⊙ Roots of plants, burrowing animals, insects and microorganisms.
- ⊙ Water relations — wetting and drying.
- ⊙ Changes in chemical composition and volume.

Physical processes primarily result in the breakdown of rocks into smaller and smaller particles. As the particles become smaller, various living organisms begin to have a great impact on soil formation because they contribute organic matter

In addition, the smaller particles speed chemical processes which result in new chemical compounds. All of these processes are greatly influenced by climate, especially temperature and precipitation.

Climate

The amount of water entering a soil influences the movement of calcium and other chemical compounds in the soil. Ultimately, if more chemicals are removed, the soils will be deeper and more developed. Precipitation influences vegetation and, therefore, greatly determines the organic matter content of soils.

Because of greater precipitation, native vegetation included luxuriant growth of the tallgrass prairie. In western Nebraska where precipitation is about half that in the east, plants of the shortgrass prairies grow much less abundantly. Thus, soil organic matter content is greater in the east than in the west.

Higher temperatures can speed the rate of organic matter decomposition. Temperatures are typically higher in the southern portion of the state than in the northern portion. Because of this trend, organic matter content decreases from north to south. However, the change in organic matter content from north to south due to temperature is minuscule when compared to the change from east to west due to precipitation.

Living Organisms

The most abundant living organism in the soil is vegetation. Vegetation influences the kind of soil developed because plants differ in their root systems, size, above ground vegetative volume, nutrient content and life cycle. Soils formed under trees are greatly different from soils formed under grass even though other soil-forming factors are similar. Trees and grass vary considerably in their search for food and water and in the amount of various chemicals taken up by roots and deposited in or on top of the soil when tree leaves and grass blades die.

Soils formed under grass are much higher in organic matter than soils formed under forests because of their massive fibrous root structure and annual senescence of above ground vegetation.

Grassland

Soils tend to be darker, particularly to greater depths, and have a more stable structure than forest soils. Soils developed under grass are generally more fertile and best suited for crop production. Nebraska soils from any parent material are nearly all formed under grass and, with adequate water, can be very productive.

The kind of plant growing influences residue composition. For example, the decay products from conifer tree needles are different from those of hardwood tree leaves. These decay products affect soil formation and development differently when water moves them through the soil.

The kind of vegetation and climate also affects the kind and numbers of other organisms that live in the soil, such as insects, small animals, and microorganisms. Organisms chew, tear and digest plant and animal material, causing it to undergo further biochemical action as it decays. Nondecomposed plant and animal material may be consumed by some organisms while others feed off of organism excrements.

There are a multitude of organisms living in the soil. Included among them are mites, snails, beetles, millipedes, springtails, worms, ground squirrels, gophers, grubs, nematodes, and microorganisms (e.g., bacteria, fungi, actinomycetes and algae). Microorganisms are the most abundant organisms in the soil.

The activity of soil organisms is strongly influenced by soil temperature, acidity and soil-water relations. Their major contributions to soil are improved soil structure, nutrient transformations and fertility, aeration and enhanced productivity.

Under forests, soil microorganisms are more diverse than under grasslands; however, microorganisms under grasslands are more active and have greater mass than under forest conditions. In general, cultivated fields have fewer organisms than virgin areas. A generalized ratio for the mass of organisms under grass/meadow:oak forest:spruce forest is 13:5:1.

Among soil organisms, bacteria are most abundant followed by actinomycete (rod-shaped microorganisms) and earthworms. As much as 4,000 pounds of bacteria can be present per acre-furrow slice (furrow slice = a 6-inch depth of soil). This is more than four times the mass of earthworms that can be present.

Because of the quantity of organisms present in the soil and their ability to accelerate the decay of organic material, they play a major role in soil formation.

Topography

Variations in topography affect moisture and temperature relations. While Nebraska is considered to be in the Great Plains, the topography within its borders varies greatly. From a broad perspective the state can be divided into regions encompassing valleys, sandhills, plains, rolling hills, dissected plains, bluffs and escarpments, and valley-side slopes.

On a local scale, we can compare a nearly level field with one that is hilly. The steeper the slope, the more influence topography has on soil development on hills and steep land. Runoff is accelerated on sloping land, so less water infiltrates the soil.

Plants, therefore, tend to have shallower root systems; and less organic matter is produced, as compared to nearly level land. Steep slopes are also subjected to more erosion which removes soil as fast, or faster, than it forms. On nearly level land, water tends to pond on the soil surface. Here, plant growth may be prolific, resulting in the production of large amounts of organic matter.

Slopes with a southern exposure are warmer and drier than slopes with a northern exposure. In fact, topography affects the micro-environment for soil formation in a manner similar to climate's affect on macro environment for soil formation.

Time

Soils have been referred to as young, mature, and old, depending on the degree of weathering. A mature soil is in equilibrium with its environment and shows full development of layers or horizons in its profile

Soils probably never reach equilibrium, but they do get older and are weathering all the time. The rate of weathering, however, slows considerably as the soil nears equilibrium with its environment.

The longer a parent material has been exposed, the greater the degree of weathering and the more developed the soil. Soils in southeast Kansas, for example, are highly weathered. Parent materials in southeast Kansas have been exposed for about 200 million years. This compares to the loess soils in Nebraska, which are only 10 to 50 thousand years old.

WEATHERING

Introduction

- ⊙ Weathering breaks down and alters rocks and minerals at or near Earth's surface and is divided into physical weathering and chemical weathering.

- ⊙ The products of weathering combine with organic material to form the soils that yield the food that sustains us, the timber that shelters us, and the fibers that clothe us.

Weathering is the name given to the process by which rocks are broken down to form soils. Rocks and geological sediments are the main parent materials of soils (the materials from which soils have formed). There is a very wide variety of rocks in the world, some acidic, some alkaline, some coarse-textured like sands, and some fine-textured and clayey.

It is from the rocks and sediments that soils inherit their particular texture. When you see rocks in the landscape it is easy to appreciate how long the process of breaking down rocks to form soil takes. In fact, it can take over 500 years to form just one centimetre of soil from some of the harder rocks. Fortunately, in some respects at least, huge amounts of rocks were broken down during the Ice Age over 10,000 years ago and converted into clays, sands or gravels, from which state it was easier to form soils.

There are three main types of weathering; physical, chemical and biological. *Physical weathering* is the influence of processes such as freezing and thawing, wetting and drying, and shrinking and swelling on rocks and other sediments, leading to their breakdown into finer and finer particles. *Chemical weathering* is the decomposition of rocks through a series of chemical processes such as acidification, dissolution and oxidation.

Some minerals, while stable within solid rock, become less stable on being more exposed to the atmosphere and so begin to alter in the rocks near the surface, destabilising the rocks. *Biological weathering* is the effect of living organisms on the break down of rock. This involves, for example, the effects of plant roots and soil organisms. Respiration of carbon dioxide by plant roots can lead to the formation of carbonic acid which can chemically attack rocks and sediments and help to turn them into soils. There are a whole range of weathering processes at work near the surface of the soil, acting together to break down rocks and minerals to form soil. These weathering processes have given rise to most of the world's soils.

A key concept to understand is how erosion, and thus soil formation, is a continual process. As rocks and sediments are eroded away, so more of the solid rock beneath becomes vulnerable in turn to weathering and breakdown. The natural processes of nature, in the form of wind, rain, snow and ice, start to have their effect on these rocks and sediments as they 'come within their range'.

Once the process starts, then other physical, chemical and biological processes also start to contribute to the breakdown of the rocks, leading to the formation of the precious soil. Most of the tiny particles making up our soils will have started as solid rock. Little or nothing will grow directly in rock; before plant life can flourish the rock first needs to be broken down to form soil. It is true to say that weathering and the formation of soil provide an excellent example of the wonders of nature.

As human beings we wage a constant battle against a silent foe, weathering, the decomposition and disintegration of features at Earth's surface. Although it has little of the power and glory of volcanoes or earthquakes, weathering acts on all features at or near Earth's surface, modifies the landscape around us, and generates perhaps our most essential resource, soil.

Different Types of Weathering

Physical Weathering

Physical weathering is the breaking of rocks into smaller pieces. This can happen through exfoliation, freeze-thaw cycles, abrasion, root expansion, and wet-dry cycles.

- **Exfoliation**: When temperature of rocks rapidly changes that can expand or crack rocks. This especially happens with granitic rocks as they were cooling, like at Yosemite National Park.

- **Freeze-thaw**: When water freezes, it expands. If moisture seeps into cracks before winter, it can then freeze, driving the rocks apart.

- **Abrasion:** When the wind blows, it can pick up sand and silt, and literally sandblast rocks into pieces.

- **Root Expansion:** Like freeze thaw, roots grow bigger every year. These roots can drive the roots apart.

The rocks are disintegrated and are broken down to comparatively smaller pieces, without producing any new substances

Physical condition of rocks

The permeability of rocks is the most important single factor. Coarse textured (porous) sand stone weather more readily than a fine textured (almost solid) basalt. Unconsolidated volcanic ash weather quickly as compared to unconsolidated coarse deposits such as gravels.

Action of Temperature

The variations in temperature exert great influence on the disintegration of rocks.

- During day time, the rocks get heated up by the sun and expand. At night, the temperature falls and the rocks get cooled and contract.

- This alternate expansion and contraction weakens the surface of the rock and crumbles it because the rocks do not conduct heat easily.

- The minerals within the rock also vary in their rate of expansion and contraction

- The cubical expansion of quartz is twice as feldspar

- Dark coloured rocks are subjected to fast changes in temperature as compared to light coloured rocks

- The differential expansion of minerals in a rock surface generates stress between the heated surface and cooled un expanded parts resulting in fragmentation of rocks.

- This process causes the surface layer to peel off from the parent mass and the rock ultimately disintegrates. This process is called Exfoliation

Action of Water

Water acts as a disintegrating, transporting and depositing agent.

Fragmentation and transport

Water beats over the surface of the rock when the rain occurs and starts flowing towards the ocean

- Moving water has the great cutting and carrying force.
- It forms gullies and ravines and carries with the suspended soil material of variable sizes.
- Transporting power of water varies. It is estimated that the transporting power of stream varies as the sixth power of its velocity i.e the greater the speed of water, more is the transporting power and carrying capacity.

Speed/Sec Carrying capacity

15 cm Fine sand

30 cm Gravel

1.2 m Stones (1kg)

9.0 m Boulders (several tons)

The disintegration is greater near the source of river than its mouth

Action of freezing

Frost is much more effective than heat in producing physical weathering

- In cold regions, the water in the cracks and crevices freezes into ice and the volume increases to one tenth
- As the freezing starts from the top there is no possibility of its upward expansion. Hence, the increase in volume creates enormous out ward pressure which breaks apart the rocks

Alternate wetting and Drying

Some natural substances increase considerably in volume on wetting and shrink on drying. (e.g.) smectite, montmorilonite

- During dry summer/ dry weather – these clays shrink considerably forming deep cracks or wide cracks.
- On subsequent wetting, it swells.
- This alternate swelling and shrinking/ wetting or drying of clay enriched rocks make them loose and eventually breaks

Action of glaciers

- In cold regions, when snow falls, it accumulates and change into a ice sheet.

- These big glaciers start moving owing to the change in temperature and/or gradient.

- On moving, these exert tremendous pressure over the rock on which they pass and carry the loose materials

- These materials get deposited on reaching the warmer regions, where its movement stops with the melting of ice

Action of wind

- Wind has an erosive and transporting effect. Often when the wind is laden with fine material viz., fine sand, silt or clay particles, it has a serious abrasive effect and the sand laden winds itch the rocks and ultimately breaks down under its force

- The dust storm may transport tons of material from one place to another. The shifting of soil causes serious wind erosion problem and may render cultivated land as degraded (e.g) Rajasthan deserts

Atmospheric electrical phenomenon

It is an important factor causing break down during rainy season and lightning breaks up rocks and or widens cracks

Chemical Weathering

Chemicals react in the environment all the time, and these cause chemical weathering. Major chemical reactions include carbonation, dissolution, hydration, hydrolysis, and oxidation-reduction reaction. All of these reactions have water involved with them.

- **Carbonation:** When water reacts with carbon dioxide, it creates carbonic acid, which can dissolve softer rocks.

- **Dissolution:** Limestone and rocks high in salt dissolve when exposed to water. The water carries away the ions.

- **Hydrolysis:** Minerals in the rock react with water and surrounding acids. The hydrogen atoms replace other cations. Feldspar hydrate to clay.

- **Oxidation-Reduction:** Water and rock particles react with oxygen. This causes the minerals and materials to rust and turn red.

If the area is hot and humid, chemical weathering is more prevalent. If it is drier, physical weathering is more predominant.

Since the chemical reactions occur largely on the surface of the rocks, therefore the smaller the fragments, the greater the surface area per unit volume available for reaction. The effectiveness of chemical weathering is closely related to the mineral composition of rocks. (e.g) quartz responds far slowly to the chemical attack than olivine or pyroxene.

Chemical Processes of weathering

Hydration Chemical combination of water molecules with a particular substance or mineral leading to a change in structure. Soil forming minerals in rocks do not contain any water and they under go hydration when exposed to humid conditions. Up on hydration there is swelling and increase in volume of minerals. The minerals loose their luster and become soft. It is one of the most common processes in nature and works with secondary minerals, such as aluminium oxide and iron oxide minerals and gypsum.

Example:

a. $2Fe_2O_3 + 3HOH \rightarrow 2Fe2O_3.3H_2O$
   (Haematite) (red)   (Limonite) (yellow)

b. $Al_2O_3 + 3HOH \rightarrow Al_2O_3.3H_2O$
   (Bauxite)   (Hyd. aluminium Oxide)

c. $CaSO_4 + 2H_2O \rightarrow CaSO_4.2H_2O$
   (Anhydrite)   (Gypsum)

d. $3(MgO.FeO.SiO_2) + 2H_2O \rightarrow 3MgO.2SiO_2.2H_2O + SiO_2 + 3H_2O$
   (Olivine)   (Serpentine)

Hydrolysis

Most important process in chemical weathering. It is due to the dissociation of H_2O into H^+ and OH^- ions which chemically combine with minerals and bring about changes, such as exchange, decomposition of crystalline structure and formation of new compounds. Water acts as a weak acid on silicate minerals.

$KAlSi_3O_8 + H_2O \rightarrow HAlSi_3O_8 + KOH$

(Orthoclase)   (Acid silt clay)

$HAlSi_3O_8 + 8HOH \rightarrow Al_2O_3.3H_2O + 6H_2SiO_3$

(recombination)   (Hyd. Alum.oxide) (Silicic acid)

This reaction is important because of two reasons

- Clay, bases and silicic acid - the substances formed in these reactions - are available to plants

- Water often containing CO_2 (absorbed from atmosphere), reacts with the minerals directly to produce insoluble clay minerals, positively charged metal ions (Ca^{++}, Mg^{++}, Na^+, K^+) and negatively charged ions (OH^-, HCO_3^-) and some soluble silica – all these ions are made available for plant growth.

Solution

Some substances present in the rocks are directly soluble in water. The soluble substances are removed by the continuous action of water and the rock no longer remains solid and form holes, rills or rough surface and ultimately falls into pieces or decomposes. The action is considerably increased when the water is acidified by the dissolution of organic and inorganic acids. (e.g) halites, NaCl

$NaCl + H_2O \rightarrow Na^+, Cl^-, H^2O$ (dissolved ions with water)

Carbonation: Carbon di oxide when dissolved in water it forms carbonic acid

$2H_2O + CO_2 \rightarrow H_2CO_3$

This carbonic acid attacks many rocks and minerals and brings them into solution. The carbonated water has an etching effect up on some rocks, especially lime stone. The removal of cement that holds sand particles together leads to their disintegration.

Components of Soils

A soil is simply a porous medium consisting of minerals, water, gases, organic matter, and microorganisms. The traditional definition is: **Soil** is a dynamic natural body having properties derived from the combined effects of **climate** and **biotic activities**, as modified by **topography**, acting on **parent materials** over **time**.

There are five basic components of soil that, when present in the proper amounts, are the backbone of all terrestrial plant ecosystems.

Mineral

The largest component of soil is the mineral portion, which makes up approximately 45% to 49% of the volume. Soil minerals are derived from two principal mineral types. **Primary minerals**, such as those found in sand and silt, are those soil materials that are similar to the parent material from which they formed. They are often round or irregular in shape. **Secondary minerals**, on the other hand, result from the weathering of the primary minerals, which releases important ions and forms more stable mineral forms such as silicate clay. Clays have a large surface area, which is important for soil chemistry and water-holding capacity. Additionally, negative and neutral charges found around soil minerals influences the soil's ability to retain important nutrients, such as cations, contributing to a soils cation exchange capacity (CEC).

Some of the key minerals include:

- **Carbon–** The building block of all organic chemistry.
- **Calcium–** In plants, calcium is a structural component of the cell wall. In plants, fungi and bacteria, calcium is a cofactor in regulating internal chemical processes. In soil, calcium shifts the pH to alkaline (vs acid).
- **Manganese–** An essential element of chlorophyll (think Photosynthesis)
- **Nitrogen–** All plant and animal proteins contain nitrogen. It is essential to life.
- **Potassium–** An essential plant nutrient involved in the regulation of photosynthesis and water management in plants.

⦿ **Phosphorous-** an essential element in the biochemistry of photosynthesis, metabolism of sugars, energy management, cell division and the transfer of genetic material. In fact, phosphorous is essential to all life forms.

⦿ **Sulfur–** An important component of amino acids and proteins.

The **texture** of a soil is based on the percentage of sand, silt, and clay found in that soil. The identification of sand, silt, and clay are made based on size. The following is used in the United States:

Sand 0.05 – 2.00 mm in diameter

Silt 0.002 – 0.05 mm in diameter

Clay < 0.002 mm in diameter

The texture of a soil can be determined from its sand, silt, and clay content using a textural triangle. The triangle (Fig 2) is the one created by the U.S. Department of Agriculture's Natural Resources Conservation Service and is primarily used in the United States. Percent clay in this triangle is read on the lefthand side of the triangle, the percent silt is read on the righthand side, and the percent sand is on the bottom. For example, if a soil contains 20% clay, 40% sand, and 40% silt (total = 100%), then it is a loam.

Water

Water is the second basic component of soil. Water can make up approximately 2% to 50% of the soil volume. Water is important for transporting nutrients to growing plants and soil organisms and for facilitating both biological and chemical decomposition. Soil water availability is the capacity of a particular soil to hold water that is available for plant use.

The capacity of a soil to hold water is largely dependent on soil texture. The more small particles in soils, the more water the soil can retain. Thus, clay soils having the greatest water-holding capacity and sands the least. Additionally, organic matter also influences the water-holding capacity of soils because of organic matter's high affinity for water. The higher the percentage of organic material in soil, the higher the soil's water-holding capacity.

The point where water is held microscopically with too much energy for a plant to extract is called the "wilting coefficient" or "permanent wilting point." When water is bound so tightly to soil particles, it is not available for most plants to extract, which limits the amount of water available for plant use. Although clay can hold the most water of all soil textures, very fine micropores on clay surfaces hold water so tightly that plants have great difficulty extracting all of it. Thus, loams and silt loams are considered some of the most productive soil textures because they hold large quantities of water that is available for plants to use.

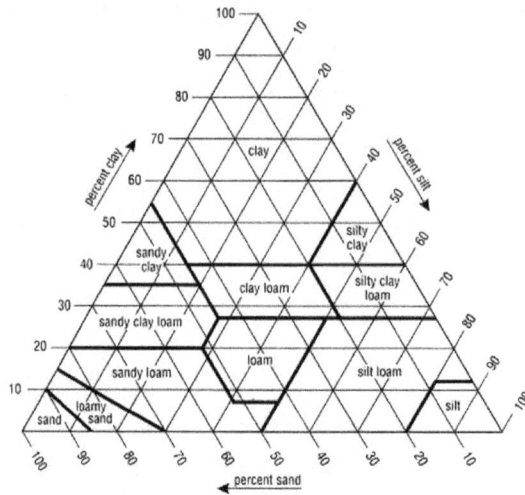

Fig. 2. The U.S. Department of Agriculture Soil Texture Triangle is used to determine the overall texture of soil based on the percentage of sand, silt, and clay.

ORGANIC MATTER

Organic matter is the next basic component that is found in soils at levels of approximately 1% to 5%. Organic matter is derived from dead plants and animals and as such has a high capacity to hold onto and/or provide the essential elements and water for plant growth. Soils that are high in organic matter also have a high CEC and are, therefore, generally some of the most productive for plant growth.

Organic matter also has a very high "plant available" water-holding capacity, which can enhance the growth potential of soils with poor water-holding capacity such as sand. Thus, the percent of decomposed organic matter in or on soils is often used as an indicator of a productive and fertile soil. Over time, however, prolonged decomposition of organic materials can lead it to become unavailable for plant use, creating what are known as recalcitrant carbon stores in soils.

GASES

Gases or air is the next basic component of soil. Because air can occupy the same spaces as water, it can make up approximately 2% to 50% of the soil volume. Oxygen is essential for root and microbe respiration, which helps support plant growth. Carbon dioxide and nitrogen also are important for belowground plant functions such as for nitrogen-fixing bacteria. If soils remain waterlogged (where gas is displaced by excess water), it can prevent root gas exchange leading to plant death, which is a common concern after floods.

MICROORGANISMS

Microorganisms are the final basic element of soils, and they are found in the soil in very high numbers but make up much less than 1% of the soil volume. A common estimate is that one thimble full of topsoil may hold more than 20,000 microbial organisms. The largest of the these organisms are earthworms and nematodes and the smallest are bacteria, actinomycetes, algae, and fungi.

Microorganisms are the primary decomposers of raw organic matter. Decomposers consume organic matter, water, and air to recycle raw organic matter into humus, which is rich in readily available plant nutrients.

Other specialized microorganisms such as nitrogen-fixing bacteria have symbiotic relationships with plants that allow plants to extract this essential nutrient. Such "nitrogen-fixing" plants are a major source of soil nitrogen and are essential for soil development over time. Mycorrhizae are fungal complexes that form mutalistic relationships with plant roots. The fungus grows into a plant's root, where the plant provides the fungus with sugar and, in return, the fungus provides the plant root with water and access to nutrients in the soil through its intricate web of hyphae spread throughout the soil matrix. Without microbes, a soil is essentially dead and can be limited in supporting plant growth.

10

SOIL WATER RETENTION: MOVEMENT OF SOIL WATER, INFILTRATION, PERCOLATION, PERMEABILITY AND DRAINAGE

INTRODUCTION

As complex as it is, soil can be described simply. It consists of four major components: air, water, organic matter, and mineral matter (Fig. 1). In an ideal **soil**, air and water fill the pore space and compose about 50 percent of the volume; organic matter accounts for about 1-5 percent of the soil volume; and mineral matter accounts for the remaining 45-49 percent. The partitioning of these four components vary considerably. For example, an organic soil in Michigan may be 45 percent organic, while a desert soil from Arizona may be 60 percent mineral.

The mineral and organic matter fractions of the soil are the solids and serve as the storehouse and exchange sites for plant nutrients and other chemicals. They are important from a fertility and environmental standpoint. It is these fractions, along with cultural practices, that influence other physical properties and processes.

MOVEMENT OF SOIL WATER, INFILTRATION, PERCOLATION, PERMEABILITY AND DRAINAGE

Introduction

As complex as it is, soil can be described simply. It consists of four major components: air, water, organic matter, and mineral matter (Fig. 1). In an ideal soil, air and water fill the pore space and compose about 50 percent of the volume; organic matter accounts for about 1-5 percent of the soil volume; and mineral matter accounts for the remaining 45-49 percent. The partitioning of these four components vary considerably. For example, an organic soil in Michigan may be 45 percent organic, while a desert soil from Arizona may be 60 percent mineral.

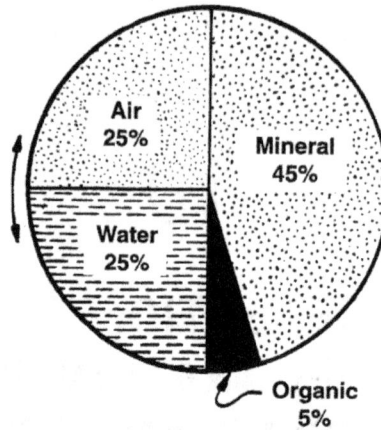

Fig 1. Major components of soils.

The mineral and organic matter fractions of the soil are the solids and serve as the storehouse and exchange sites for plant nutrients and other chemicals. They are important from a fertility and environmental standpoint. It is these fractions, along with cultural practices, that influence other physical properties and processes.

Retention of Water by Soil

The soils hold water (moisture) due to their colloidal properties and aggregation qualities. The water is held on the surface of the colloids and other particles and in the pores. The forces responsible for retention of water in the soil after the drainage has stopped are due to surface tension and surface attraction and are called surface moisture tension. This refers to the energy concept in moisture retention relationships. The force with which water is held is also termed as suction.

Water retention in soil can be understood as the water retained by the soil after it runs through the soil pores to join water bodies such as groundwater or surface streams. Pores in the soil can be defined as the air-spaces that exist in between soil particles.

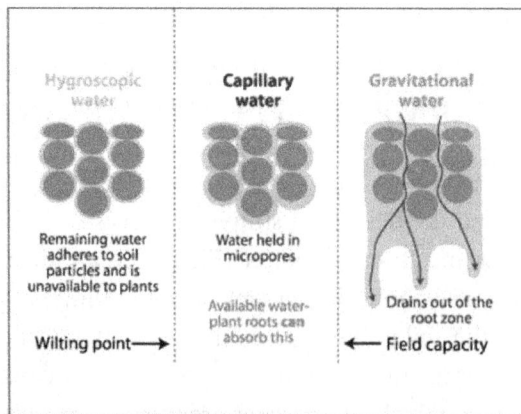

Fig. 2. Available water for plant growth

Water retention is mainly dependant on the particle size of the soil. The finer the soil particles, the higher the chance that water molecules shall hold on to soil particles, such as in clay, as opposed to sandy soil, that has large and coarse particles that are not cohesive.

The water retention by soil is critical for plants and acts as the chief source of moisture for it in almost all habitats. Other than percolation through the soil, soil moisture can also deplete due to evaporation directly from the soil and by transpiration by plants.

SOIL AGE

The weathering forces, as influenced by the five soil-forming factors, result in the formation of soil from rocks and parent materials. However, the soil formed is not just a loose unconsolidated, disorganized mass of various size particles. The soil has definite organized qualitative and quantitative characteristics reflecting the forces from which it was formed.

The age of most persons are measured in years since birth. For people and soils, there are also ways to judge age in terms of "maturity". The age of a soil is determined by the amount of weathering that has occurred; that is, to what extent the parent material has been converted to distinct horizons or soil layers. Soil age is based on three general criteria:

- ⊙ The more horizons that are present, the older the soil.
- ⊙ The thicker the horizons, the older the soil.
- ⊙ The more difference there is between adjacent horizons, the older the soil.

Soil age is not an exact measure. It is usually described simply as young, mature or old. A young soil has a thin A horizon and often no B horizon. A mature soil has A and B horizons of average thickness which show some differences from each other. An old soil shows thick horizons which are very different from each other.

As based on the size of the soil particles, there are four classificatory systems for the identification of soils -

1. US Bureau of Soil Classification
2. International Classification
3. M.I.T Classification
4. Indian Standard Classification

The Indian Standard classificatory system was formulated originally for the classification of soils primarily for engineering purposes. This is because the draft for soil classification was prepared by the Soil Engineering Sectional Committee and was approved by the Civil Engineering Division Council. The final draft was adopted by the Bureau of Indian Standards on December 19th, 1970. This system divides soils into three broad categories based on the properties of soil particles (Bureau of Indian Standards, 2004) -

i. **Clay** - the particles are microscopic to sub-microscopic and exhibit plasticity, allowing it to retain the most water.

ii. **Silt** - the particles are fine grains, but exhibit less plasticity, making this form retain lesser water.

iii. **Sand and Gravel** - aggregates of comparably larger particles that are coarse and loosely bound thus lacking cohesion. The least water retention is possible in this form of soil.

Different topographic and climactic patterns result in varied behaviour of soils and thus require a variety of approaches to analyze and implement soil management techniques for water retention. Soil can sometimes pose problems for not being as desired, and these problems can broadly be grouped under chemical and physical problems.

Chemical problems include high salinity or acidity in soils, along with the presence of other toxic chemicals such as phosphorous in soil (NAU, 2013). This problem becomes especially pertinent in agriculture where crop yield or productivity could dwindle due to chemicals used in the agricultural process such as pesticides and herbicides. Among a large gamut of solutions and applications, the most common one is the use of ecologically beneficial green manure. Agricultural soil should also be frequently and properly drained to achieve effects such as the leaching of saline moisture in soils.

The physical problems can involve soil that is not able to contain much water due to lack of cohesion or due to a rigidity that can occur owing to encrustation, or a very clayey surface. Shallow depth of soil, soil that is too clayey, or the presence of hard opaque surfaces underneath can also present problems to water retention and there can be water-logging when too much water is added to soil. These require artificial solutions to soil management that frequently involves the mixing of soil with other different forms of soil. Incorporating organic matter and regulating drainage are also frequently applied solutions.

There are various methods to enhancing the water retention capacity of soil. Some methods are more traditional, and also conventional, while some involve the utilization of technology. While most of technological investment regarding water retention in soils involves technologies for enumeration and generation of data, technological solutions can vary from simple, affordable, everyday solutions to solutions utilizing high-end technology.

Some of the simple solutions include application of organic solutions such as drought resistant crop varieties and organisms that increase the fertility of soil, management and design of irrigation according to soil properties, application of biochar - produced from biomass for low-cost carbon sequestration in soil - making soil less porous, use of the roots of plants that grip soil, and application of natural by-products such as poultry litter that provide greater cohesiveness to soil. The solutions can also range towards using complex technologies such as mapping the global water cycle in relation to water retention in soil, and preparation of dietary fibres that have high water-holding capacity from food sources used in soil. There is however, a leaning in technological progress in engineering water retention in soil to introduce organic elements in the soil instead of inorganic matter.

The water retained in the soil by following ways

Cohesion and adhesion forces

These two basic forces are responsible for water retention in the soil. One is the attraction of molecules for each other i.e., cohesion. The other is the attraction of water molecules for the solid surface of soil i.e. adhesion. By adhesion, solids (soil) hold water molecules rigidly at their soil - water interfaces. These water molecules in turn hold by cohesion. Together, these forces make it possible for the soil solids to retain water.

Surface tension

This phenomenon is commonly evidenced at water- air interfaces. Water behaves as if its surface is covered with a stretched elastic membrane. At the surface, the attraction of the air for the water molecules is much less than that of water molecules for each other. Consequently, there is a net downward force on the surface molecules, resulting in sort of a compressed film (membrane) at the surface. This phenomenon is called surface tension.

Polarity or dipole character

The retention of water molecules on the surface of clay micelle is based on the dipole character of the molecule of water. The water molecules are held by electrostatic force that exists on the surface of colloidal particles. By virtue of their dipole character and under the influence of electrostatic forces, the molecules of water get oriented (arranged) on the surface of the clay particles in a particular manner.

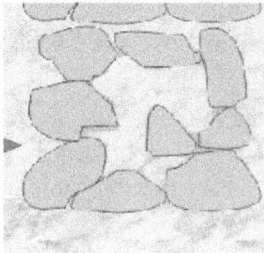

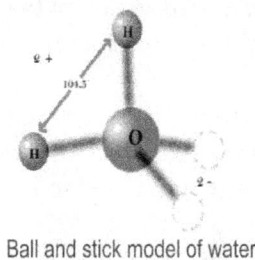

Ball and stick model of water

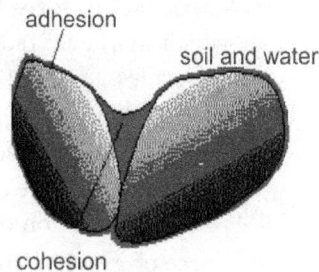

Fig. 3. Cohesion

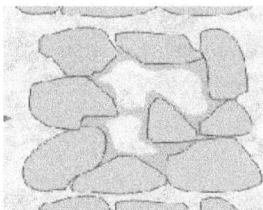

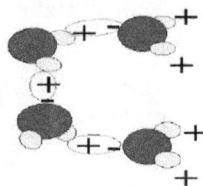

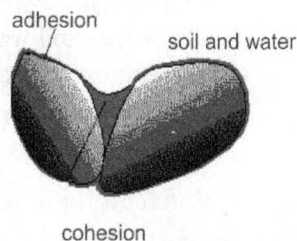

Fig. 4. Cohesion

Each water molecule carries both negative and positive charges. The clay particle is negatively charged. The positive end of water molecule gets attached to the negatively charged surface of clay and leaving its negative end outward. The water molecules attached to the clay surface in this way present a layer of negative charges to which another layer of oriented water molecules is attached. The number of successive molecular layers goes on increasing as long as the water molecules oriented. As the molecular layer gets thicker, orientation becomes weaker, and at a certain distance from the particle surface the water molecules cease to orientate and capillary water (liquid water) begins to appear.

Due to the forces of adsorption (attraction) exerted by the surface of soil particles, water gets attached on the soil surface. The force of gravity also acts simultaneously, which tries to pull it downwards. The surface force is far greater than the force of gravity so water may remain attached to the soil particle. The water remains attached to the soil particle or move downward into the lower layers, depending on the magnitudeof the resultant force.

Potentials

Soil water potential

The retention and movement of water in soils, its uptake and translocation in plants and its loss to the atmosphere are all energy related phenomenon. The more strongly water is held in the soil the greater is the heat (energy) required. In other words, if water is to be removed from a moist soil, work has to be done against adsorptive forces. Conversely, when water is adsorbed by the soil, a negative amount of work is done. The movement is from a zone where the free energy of water is high (standing water table) to one where the free energy is low (a dry soil). This is called soil water energy concept.

Free energy of soil solids for water is affected by:

    i. Matric (solid) force i.e., the attraction of the soil solids for water (adsorption) which markedly reduces the free energy (movement) of the adsorbed water molecules.

    ii. Osmotic force i.e., the attraction of ions and other solutes for water to reduce the free energy of soil solution. Matric and Osmotic potentials are negative and reduce the free energy level of the soil water. These negative potentials are referred as suction or tension.

    iii. Force of gravity: This acts on soil water, the attraction is towards the earth's center, which tends to pull the water down ward. This force is always positive. The difference between the energy states of soil water and pure free water is known as soil water potential. Total water potential (P_t) is the sum of the contributions of gravitational potential (P_g), matric potential (P_m) and the Osmotic potential or solute potential (P_o).

$$P_t = P_g + P_m + P_o$$

Potential represents the difference in free energy levels of pure water and of soil water. The soil water is affected by the force of gravity, presence of soil solid (matric) and of solutes.

SOIL MOISTURE CONSTANTS

Earlier classification divided soil water into gravitational, capillary and hygroscopic water. The hygroscopic and capillary waters are in equilibrium with the soil under given condition.

The hygroscopic coefficient and the maximum capillary capacity are the two equilibrium points when the soil contains the maximum amount of hygroscopic and capillary waters, respectively. The amount of water that a soil contains at each of these equilibrium points is known as soil moisture constant.

The soil moisture constant, therefore, represents definite soil moisture relationship and retention of soil moisture in the field

The three classes of water (gravitational, capillary and hygroscopic) are however very broad and do not represent accurately the soil - water relationships that exists under field conditions.

Though the maximum capillary capacity represents the maximum amount of capillary water that a soil holds, the whole of capillary water is not available for the use of the plants. A part of it, at its lower limit approaching the hygroscopic coefficient is not utilized by the plants. Similarly a part of the capillary water at its upper limit is also not available for the use of plants. Hence two more soil constants, viz., field capacity and wilting coefficient have been introduced to express the soil-plant-water relationships as it is found to exist under field conditions.

1. **Field capacity:** Assume that water is applied to the surface of a soil. With the downward movement of water all macro and micro pores are filled up. The soil is said to be saturated with respect to water and is at maximum water holding capacity or maximum retentive capacity. It is the amount of water held in the soil when all pores are filled.Sometimes, after application of water in the soil all the gravitational water is drained away, and then the wet soil is almost uniformly moist. The amount of water held by the soil at this stage is known as the field capacity or normal moisture capacity of that soil. It is the capacity of the soil to retain water against the downward pull of the force of gravity. At this stage only micropores or capillary pores are filled with water and plants absorb water for their use. At field capacity water is held with a force of 1/3 atmosphere. Water at field capacity is readily available to plants and microorganism.

2. **Wilting coefficient:** As the moisture content falls, a point is reached when the water is so firmly held by the soil particles that plant roots are unable to draw it. The plant begins to wilt. At this stage even if the plant is kept in a saturated atmosphere it does not regain its turgidity and wilts unless water is applied to the soil. The stage at which this occurs is termed the Wilting point and the percentage amount of water held by the soil at this stage is known as the Wilting Coefficient. It represents the point at which the soil is unable to supply water to the plant. Water at wilting coefficient is held with a force of 15 atmosphere.

3. **Hygroscopic coefficient:** The hygroscopic coefficient is the maximum amount of hygroscopic water absorbed by 100 g of dry soil under standard conditions of humidity (50% relative humidity) and temperature (15°C). This tension is equal to a force of 31 atmospheres. Water at this tension is not available to plant but may be available to certain bacteria.

4. **Available water capacity:** The amount of water required to apply to a soil at the wilting point to reach the field capacity is called the "available" water. The water supplying power of soils is related to the amount of available water a soil can hold. The available water is the difference in the amount of water at field capacity (- 0.3 bar) and the amount of water at the permanent wilting point (- 15 bars).

5. **Maximum water holding capacity:** It is also known as maximum retentive capacity. It is the amount of moisture in a soil when its pore spaces both micro and macro capillary are completely filled with water. It is a rough measure of total pore space of soil. Soil moisture tension is very low between 1/100th to 1/1000th of an atmosphere or pF 1 to 0.

6. **Sticky point moisture:** It represents the moisture content of soil at which it no longer sticks to a foreign object. The sticky point represents the maximum moisture content at which a soil remains friable. Sticky point moisture values vary nearly approximate to the moisture equivalent of soils. Summary of the soil moisture constants, type of water and force with which it held is given in following table.

Table: Soil moisture constants and range of tension and pF

S.No.	Moisture class	Tension (atm)	pF
1	Chemically combined	Very high	---
2	Water vapour	Held at saturation point in the soil air	---
3	Hygroscopic	31 to 10,000	4.50 to 7.00
4	Hygroscopic coefficient	31	4.50
5	Wilting point	15	4.20
6	Capillary	1/3 to 31	2.54 to 4.50
	Moisture equivalent	1/3 to 1	2.70 to 3.00
	Field capacity	1/3	2.54
	Sticky point	1/3 (more or less)	2.54
	Gravitational	Zero or less than 1/3	<2.54
	Maximum water holding capacity	Almost zero	---

Soil water capacity

Moisture equivalent : It is defined as the percentage of water held by one centimeter thick moist layer of soil subjected to a centrifugal force of 1000 times of gravity for half an hour.

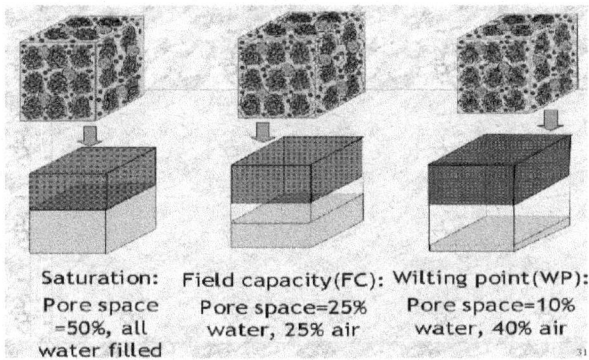

Fig. 5. Soil water capacity

FEATURES OF SOIL WATER RETENTION CURVES

The most fundamental concept to understand about soil water retention is that soil water content is positively related to soil matric potential. As soil water content decreases, matric potential also decreases, becoming more negative. When all the pores in a soil are filled with water, the soil is at its saturated water content (θ_s) and the matric potential is 0. Consider the water retention curve for the Rothamsted loam shown in Fig. 6. The intersection of the solid curve with the left-hand y-axis shows that for this soil θ_s is approximately 0.51 cm³ cm⁻³.

As we move to the right along the solid curve, we are moving toward more negative values of matric potential. The absolute value of matric potential, rather than matric potential itself, is plotted on the x-axis in this figure, as is common for water retention curve plots. The absolute value of matric potential is sometimes called suction. Using the absolute value for matric potential allows us to use a logarithmic scale for matric potential to compensate for its large numerical range relative to that of soil water content.

The water retention curve for the Rothamsted loam is flat between 100 cm (i.e. 1 cm) and approximately 102 cm (100 cm), then at lower matric potentials the curve bends downward. The highest matric potential at which air has displaced water in some of the pores of a previously saturated soil is called the air-entry potential (ψ_e). For this Rothamsted loam the air-entry potential was estimated to be -128 cm of water.

As we follow the water retention curve toward the right from the air-entry potential, we encounter a region where the decrease in water content is relatively large for each corresponding decrease in matric potential. There is a subtle inflection point approximately halfway down the descending limb of the water retention curve where the shape changes from concave to convex. The location of this inflection point may have some practical significance for soil management. The water content at this inflection point may be the optimum water content for tillage, resulting in the greatest proportion of small aggregates, and the slope of the curve at the inflection point may be a useful indicator of soil quality.

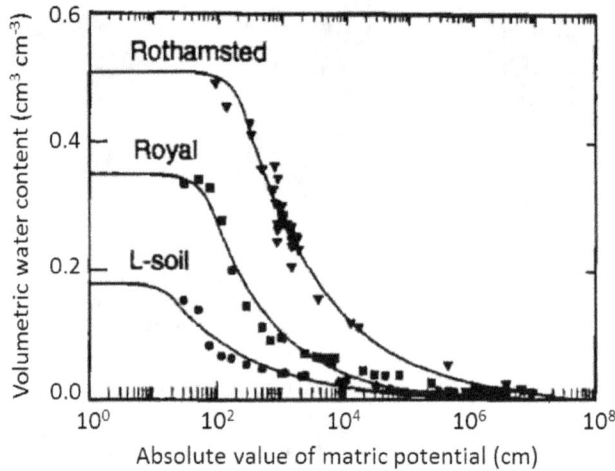

Fig. 6. Water retention curve for the Rothamsted loam

To the right of the inflection point, the steep portion of the curve tapers off into a relatively flat portion of the curve when the matric potential takes on large negative values. In this tail of the water retention curve, large decreases in matric potential are associated with only small decreases in soil water content.

Soil Properties Affecting Soil Water Retention

Another fundamental characteristic of soil water retention curves is that coarse-textured soils retain less water than fine textured soils at the same matric potential. Consider the substantial differences in the curves for the sand (L-soil), sandy loam (Royal), and loam (Rothamsted) textured soils in Fig. 6. The sand exhibits a much lower saturated water content than the loam, in this case 0.18 cm³ cm⁻³ versus 0.51 cm3 cm-3. The sand also has a higher (less negative) air-entry potential than the loam, -32 cm versus -128 cm. The water retention for the medium-textured sandy loam soil is intermediate between those of the other two soils. Throughout the subsequent chapters, one common theme will be how these substantial differences in water retention between different soil textures dramatically influence water movement, plant growth, and related processes in both managed and natural ecosystems.

A secondary influence on soil water retention is the soil bulk density (Fig. 7). If you compare compacted and un-compacted samples of the same soil, the compacted soil will typically have a lower porosity, lower saturated water content, and lower air-entry potential. Sufficiently compacted soils can also have higher water contents for matric potentials below the air-entry potential than a similar un-compacted soil. This pattern is evident for the samples with the highest bulk density in Fig. 7.

Advocates for conservation tillage, cover crops, soil quality, and, more recently, soil health have often stated that increasing soil organic matter improves soil water retention. However, the scientific evidence for this claim is somewhat unclear. While a number of studies have found that increasing organic matter increases soil water retention, a similar number of studies have found no such effect. One plausible hypothesis is that in some soils increasing organic matter results in decreased bulk density, leading indirectly to positive effects on water retention similar to those shown in Fig. 7.

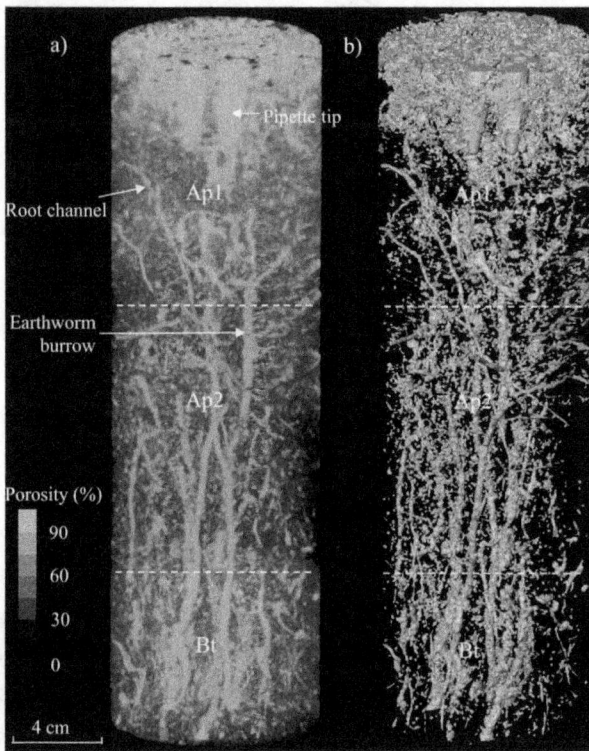

Fig. 7 A secondary influence on soil water retention is the soil bulk density

Hysteresis in Soil Water Retention

The soil water retention curve can also be influenced by whether the soil is undergoing wetting (sorption) or drying (desorption). When the soil water retention curve differs between wetting and drying, that phenomenon is called hysteresis. This phenomenon has a number of important effects on soil water dynamics. For example, hysteresis in the water retention curve can increase the amount of water that is stored near the soil surface after an infiltration and drainage event. Hysteresis can also slow the rate of solute leaching in soil under natural rainfall conditions with greater effects in coarse-textured than fine-textured soils. In subsequent chapters, we will further consider the effects of hysteresis. For now, we will examine its nature and causes.

For a soil exhibiting hysteresis, the equilibrium water content associated with any particular matric potential will be lower for a wetting curve than for a drying curve (Fig. 8). The initial water content for the wetting or drying process also plays a role. Notice in Fig. 8 the clear difference in the drying curve for the silty clay loam soil when the drying process began from full saturation compared to when the drying process began at a lower water content indicated by the point labeled "B".

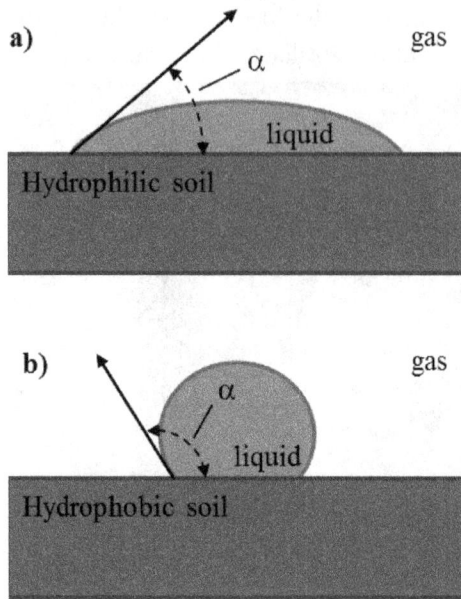

Fig. 8

Hysteresis in the soil water retention curve has multiple possible causes including: air entrapment, contact angle hysteresis, and the "ink bottle" effect. Air-entrapment occurs when a partially-drained soil is rewetted and small pockets of air become trapped in the interior pore spaces. This entrapped air cannot easily be removed, even if the soil is submerged underwater. As a result, higher water contents occur along the primary drainage curve from a fully saturated condition than those that occur during subsequent re-wetting (e.g. Fig. 8). Due to air-entrapment during re-wetting, the soil water content approaches a maximum value below the true saturated water content and this lower value is sometimes called the satiated water content. Soil chemical, physical, and biological processes can alter the amount and distribution of entrapped air over time, so the impact of air-entrapment on soil water retention can change with each subsequent re-wetting cycle.

A second potential cause of hysteresis in the soil water retention curve is a phenomenon known as contact angle hysteresis. The contact angle is the angle at which a liquid-gas interface meets a solid surface. In our context, this means the angle at which the interface between the soil solution and the soil gas phase contacts the soil solids. Mineral soils often have contact angles <90° and are classified as hydrophilic, i.e. having affinity for water. Organic soils and mineral soils in which much of the surface area becomes covered with organic coatings can have contact angles >90°, making them hydrophobic, i.e. tending to repel water.

To visualize contact angle hysteresis and how it may affect soil water retention, a thought experiment may help. Imagine if we added a sufficiently small volume of liquid to the drop in Fig. 8a, the edge of the drop would not move but the contact angle would increase slightly. Likewise if we removed a sufficiently small amount of liquid, the contact angle would decrease slightly. Thus, contact angles for wetting and drying processes are different, i.e. contact angles exhibit hysteresis. The larger contact angles during wetting versus drying lead to higher (less negative) pressure potentials for the same water contents, consistent with Fig. 8.

A third potential cause for hysteresis is the ink bottle effect, which refers to the way in which drainage from a relatively large cavity, such as the body of an old-fashioned ink bottle, can be restricted if the fluid must drain through a relatively narrow opening, such as the neck of an inverted ink bottle. The analogy is somewhat helpful, but to better understand how this phenomenon influences soil water retention, we need to understand an important related phenomenon called capillary rise. Capillary rise is the rise of liquid against the force of gravity due to the upward force produced by the attraction of the liquid molecules to a solid surface and to each other.

When you insert a small diameter tube, or capillary, into a fluid, such as water, the surface of the fluid inside the capillary may rise above that of the surrounding fluid, and the height (h) of this capillary rise is described by:

$$h = \frac{2\gamma\cos\alpha}{\rho g r} \qquad \text{(Eq. 1)}$$

Where γ is the surface tension of the fluid (N m^{-1}), α is the contact angle of the liquid-gas interface on the wall of the tube, ρ is the fluid density (kg m^{-3}), g is the acceleration due to gravity (m s^{-2}), and r is the radius of the capillary (m). Thus, the smaller the radius of the capillary, the greater the height of the capillary rise. To better understand this equation, watch this video. The pressure potential just below the capillary meniscus is simply the negative of the capillary rise.

In Fig. 9, two capillary tubes have been inserted into water. The height of the resulting capillary rise was greater for the uniformly narrow tube on the right than for the non-uniform tube on the left. During this filling or wetting phase, capillary rise could only raise water to the bottom of the tube section with the enlarged diameter. If instead both tubes had drained from an initially water filled condition, then the enlarged section would have remained water-filled and height of water in both tubes would have been equal. Thus, for capillary tubes or soil pores with non-uniform radii, that non-uniformity can cause hysteresis in the water retention curve.

Measuring Soil Water Retention Curves

Because of the complexity of soil pore networks, we are currently unable to theoretically predict soil water retention curves from first principles, although progress has been made and is being made toward that goal. Until that goal is achieved, we will continue to determine soil water retention curves primarily by empirical methods, i.e. methods based on measurements and experience rather than theory or logical reasoning.

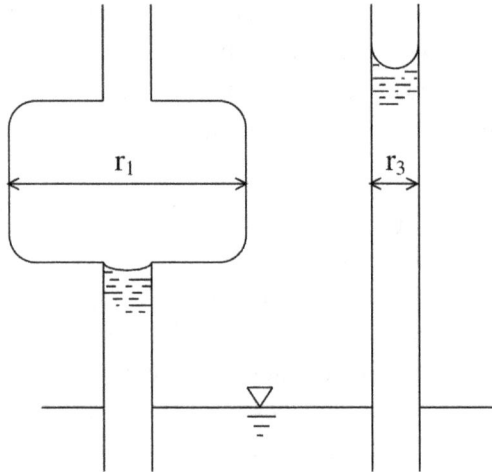

Fig. 9. Capillary tubes

Measurements of soil water retention are typically, but not always, performed in the laboratory with different methods being suitable for different portions of the possible range in soil matric potential. Near saturation, intact soil samples should be used because the soil structure and inter-aggregate pores can strongly influence water retention. At matric potentials below approximately -15 kPa, the effects of soil structure on water retention appear to be negligible and smaller homogenized soil samples are typically used.

For matric potentials between 0 kPa and approximately -10 kPa, a simple hanging water column or tension table is often used to precisely control a sample's matric potential. When the sample reaches equilibrium with the imposed matric potential, i.e. when water stops flowing, the water content of the sample can be determined by the change in the mass of the sample.

For matric potentials between -10 and -100 kPa, small pressurized chambers often called Tempe cells work well, particularly for intact soil samples. A special porous ceramic plate at the bottom of the chamber, when saturated, allows water, but not air, to flow out of the chamber. The air pressure is increased to the absolute value of the desired matric potential, and once equilibrium is reached, the water content of the sample is determined based on the volume of water which flowed out of the sample or the change in mass of the sample.

For matric potentials between -100 and -1500 kPa, specialized pressure plates in larger chambers have often been used. The principle of operation is similar to that of Tempe cells, but smaller samples of homogenized soil are used with each chamber housing multiple samples, and sometimes even multiple pressure plates. At these low matric potentials, true equilibrium may take many weeks or may never be reached, and a growing body of research suggests that data from pressure plate measurements may be unreliable at matric potentials below -100 kPa. Dewpoint potentiometers offer one alternative measurement approach in this matric potential range.

Mathematical Functions for Soil Water Retention

Once we have measured soil water retention at several values of matric potential, we often need to fit a mathematical function to the measurements to allow calculation of water content for all other possible values of matric potential. One of the earliest widely-used water retention functions, defined by:

$$\frac{\Theta - \Theta_r}{\Theta_s - \Theta_r} = \left(\frac{\Psi_e}{\Psi_m} \right)^2 \qquad \text{(Eq. 2)}$$

for $\Psi_m < \Psi_e$

Where Θ_r is the residual water content, which is conceptually the water content below which liquid water flow in the soil is no longer possible, Ψ_e is the air-entry potential, and λ is a number related to the pore size distribution of the soil. Larger values of l indicate more uniformly-sized pores, while small values indicate a wide distribution of pore sizes are present.

The most accurate way to estimate the parameters needed for these water retention functions is to obtain measurements of soil water retention across a broad range of matric potentials and then to adjust the parameters to achieve the best possible agreement with the measured values. Measured water retention curves for a loamy sand and a silt loam soil along with best-fits of the Brooks and Corey, Campbell, and van Genuchten water retention functions. All three functions fit the data reasonably, with the primary difference in this case being the sharp drop in water content at the air-entry potential predicted by the Brooks and Corey and the Campbell functions. The optimized parameters for each function are listed in

INFILTRATION, PERCOLATION, PERMEABILITY

Movement of soil water - Infiltration, percolation, permeability - Drainage

Soil acts as a sponge to take up and retain water. Movement of water into soil is called infiltration, and the downward movement of water within the soil is called percolation, permeability or hydraulic conductivity. Pore space in soil is the conduit that allows water to infiltrate and percolate. It also serves as the storage compartment for water.

Infiltration rates can be near zero for very clayey and compacted soils, or more than 10 inches per hour for sandy and well aggregated soils. Low infiltration rates lead to ponding on nearly level ground and runoff on sloping ground. Organic matter, especially crop residue and decaying roots, promotes aggregation so that larger soil pores develop, allowing water to infiltrate more readily.

Permeability also varies with soil texture and structure. Permeability is generally rated from very rapid to very slow (Table below). This is the mechanism by which water reaches the subsoil and rooting zone of plants. It also refers to the movement of water below the root zone. Water that percolates deep in the soil may reach a perched water-table or groundwater aquifer. If the percolating water carries chemicals such as nitrates or pesticides, these water reservoirs may become contaminated.

Table Permeability classification system

Permeability class	Rate (inches/hour)
Very rapid	Greater than 10
Rapid	5 To 10
Moderately rapid	2.5 To 5
Moderate	0.8 To 2.5
Moderately slow	0.2 To 0.8
Slow	0.05 To 0.2
Very slow	Less than 0.05

Infiltration and permeability describe the manner by which water moves into and through soil. Water held in a soil is described by the term water content. Water content can be quantified on both a gravimetric (g water/g soil) and volumetric (ml water/ml soil) basis. The volumetric expression of water content is used most often. Since 1 gram of water is equal to 1 milliliter of water, we can easily determine the weight of water and immediately know its volume. The following discussion will consider water content on a volumetric basis.

Saturation is the soil water content when all pores are filled with water. The water content in the soil at saturation is equal to the percent porosity. Field capacity is the soil water content after the soil has been saturated and allowed to drain freely for about 24 to 48 hours. Free drainage occurs because of the force of gravity pulling on the water.

When water stops draining, we know that the remaining water is held in the soil with a force greater than that of gravity. Permanent wilting point is the soil water content when plants have extracted all the water they can. At the permanent wilting point, a plant will wilt and not recover. Unavailable water is the soil water content that is strongly attached to soil particles and aggregates, and cannot be extracted by plants. This water is held as films coating soil particles. These terms illustrate soil from its wettest condition to its driest condition.

Several terms are used to describe the water held between these different water contents. Gravitational water refers to the amount of water held by the soil between saturation and field capacity. Water holding capacity refers to the amount of water held between field capacity and wilting point. Plant available water is that portion of the water holding capacity that can be absorbed by a plant. As a general rule, plant available water is considered to be 50 percent of the water holding capacity.

The volumetric water content measured is the total amount of water held in a given soil volume at a given time. It includes all water that may be present including gravitational, available and unavailable water.

The relationship between these different physical states of water in soil can be easily illustrated using a sponge. A sponge is just like the soil because it has solid and pore space. Obtain a sponge about 6 x 3 x 1/2 inch in size. Place it under water in a dishpan, and allow it to soak up as much water as possible. At this point, the sponge is at saturation. Now, carefully support the sponge with both hands and lift it out of the water. When the sponge stops draining, it is at field capacity, and the water that has freely drained out is gravitational water. Now, squeeze the sponge until no more water comes out. The sponge is now at permanent wilting point, and the water that was squeezed out of the sponge is the water holding capacity. About half of this water can be considered as plant available water. You may notice that you can still feel water in the sponge. This is the unavailable water.

Water in the form of precipitation or irrigation infiltrates the soil surface. All pores at the soil surface are filled with water before water can begin to move downward. During infiltration, water moves downward from the saturated zone to the unsaturated zone.

The interface between these two zones is called the wetting front. When precipitation or irrigation cease, gravitational water will continue to percolate until field capacity is reached. Water first percolates through the large pores between soil particles and aggregates and then into the smaller pores.

Available water is held in soil pores by forces that depend on the size of the pore and the surface tension of water. The closer together soil particles or aggregates are, the smaller the pores and the stronger the force holding water in the soil.

Because the water in large pores is held with little force, it drains most readily. Likewise, plants absorb soil water from the larger pores first because it takes less energy to pull water from large pores than from small pores.

Use of soil water estimates on a percentage volume basis does not allow for any practical interpretation. Therefore, water is usually converted from a percentage volume basis to a depth basis of inches of water/foot of soil.

Table Estimated soil water for three soil textures

	Inches of water/foot of soil		
	Sand	Loam	Silty clay loam
Saturation	5.2	5.8	6.1
Field capacity	2.1	3.8	4.4
Permanent wilting point	1.1	1.8	2.6

	Inches of water/foot of soil		
	Sand	Loam	Silty clay loam
Oven dry	0	0	0
Gravitational	3.1	2	1.7
Water holding capacity	1	2	1.8
Plant available	0.5	1	0.9
Unavailable	1.1	1.8	2.6

The table values are derived from laboratory analysis of soil samples. Some of this information is also published in the Soil Survey. Other techniques have been developed to estimate soil water if laboratory data is not available. Generally, field capacity is considered to be 50 percent of saturation and permanent wilting point is 50 percent of field capacity.

Water holding capacity designates the ability of a soil to hold water. It is useful information for irrigation scheduling, crop selection, groundwater contamination considerations, estimating runoff and determining when plants will become stressed. Available water capacity varies by soil texture (Table).

Table: Range of available water capacity for different soil textures

Textural class	Available water capacity, inches/foot of soil
Course sand	0.25 - 0.75
Fine sand	0.75 - 1.00
Loamy sand	1.10 - 1.20
Sandy loam	1.25 - 1.40
Fine sandy loam	1.50 - 2.00
Silt	
Loam	2.00 - 2.50
Silty clay loam	1.80 - 2.00
Silty clay	1.50 - 1.70
Clay	1.20 - 1.50

Medium textured soils (fine sandy loam, silt loam and silty clay loam) have the highest water holding capacity, while coarse soils (sand, loamy sand and sandy loam) have the lowest water holding capacity. Medium textured soils with a blend of silt, clay and sand particles and good aggregation provide a large number of pores that hold water against gravity. Coarse soils are dominated by sand and have very little silt and clay. Because of this, there is little aggregation and few small pores that will hold water against gravity. Fine textured clayey soils have a lot of small pores that hold much water against gravity. Water is held very tightly in the small pores making it difficult for plants to adsorb it.

Since soil texture varies by depth, so does water holding capacity. A soil may have a clayey surface with a silty B horizon and a sandy C horizon. To determine water holding capacity for the soil profile, the depth of each horizon is multiplied by the available water for that soil texture, and then the values for the different horizons are added together.

Water relations are greatly affected by cultural practices, but the effect is largely indirect. For instance, tillage breaks down aggregates, decreasing the number of large pores. This would cause a decrease in infiltration rate and percolation, the water content at field capacity would increase, and gravitational water would decrease. If compaction causes an increase in the number of very small pores, unavailable water may increase, and water holding capacity may decrease. As a result, the amount of plant available water would also decrease.

Methods of determination of soil moisture -

Soil Water Movement

> i. Saturated Flow
> ii. Unsaturated Flow
> iii. Water Vapour Movement

Saturated flow

This occurs when the soil pores are completely filled with water. This water moves at water potentials larger than - 33 kPa. Saturated flow is water flow caused by gravity's pull. It begins with infiltration, which is water movement into soil when rain or irrigation water is on the soil surface. When the soil profile is wetted, the movement of more water flowing through the wetted soil is termed percolation.

Hydraulic conductivity can be expressed mathematically as

$V = kf$

Where,

V = Total volume of water moved per unit time

f = Water moving force

k = Hydraulic conductivity of soil

Factors affecting movement of water

1. Texture, 2. Structure, 3. Amount of organic matter, 4. Depth of soil to hard pan, 5. Amount of water in the soil, 6. temperature and 7. Pressure

Vertical water flow

The vertical water flow rate through soil is given by Darcy's law. The law states that the rate of flow of liquid or flux through a porous medium is proportional to the hydraulic gradient in the direction of floe of the liquid (dw) At.

$QW = - k (dw) At / Ds$

Where,

QW = Quantity of water in cm^{-1}

k = rate constant (cm/s)

dw = Water height (head), cm

A = Soil area (cm2)

t = Time

ds = Soil depth (cm)

Unsaturated Flow

It is flow of water held with water potentials lower than- 1/3 bar. Water will move toward the region of lower potential (towards the greater "pulling" force). In a uniform soil this means that water moves from wetter to drier areas. The water movement may be in any direction.

The rate of flow is greater as the water potential gradient (the difference in potential between wet and dry) increases and as the size of water filled pores also increases. The two forces responsible for this movement are the attraction of soil solids for water (adhesion) and capillarity.

Under field conditions this movement occurs when the soil macropores (noncapillary) pores with filled with air and the micropores (capillary) pores with water and partly with air.

Factors Affecting the Unsaturated Flow

Unsaturated flow is also affected in a similar way to that of saturated flow. Amount of moisture in the soil affects the unsaturated flow. The higher the percentage of water in the moist soil, the greater is the suction gradient and the more rapid is the delivery.

Water Vapour Movement

The movement of water vapour from soils takes place in two ways: (a) Internal movement-the change from the liquid to the vapour state takes place within the soil, that is, in the soil pores and (b) External movement-the phenomenon occurs at the land surface and the resulting vapour is lost to the atmosphere by diffusion and convection.

The movement of water vapour through the diffusion mechanism taken place from one area to other soil area depending on the vapour pressure gradient (moving force). This gradient is simply the difference in vapour pressure of two points a unit distance apart. The greater this difference, the more rapid the diffusion and the greater is the transfer of water vapour during a unit period.

SOIL CONDITIONS AFFECTING WATER VAPOUR MOVEMENT:

There are mainly two soil conditions that affect the water vapour movement namely moisture regimes and thermal regimes. In addition to these, the various other factors which influence the moisture and thermal regimes of the soil like organic matter, vegetative cover, soil colour etc. also affect the movement of water vapour. The movement takes place from moist soil having high vapour pressure to a dry soil (low vapour pressure). Similarly the movement takes place from warmer soil regions to cooler soil region.In dry soils some water movement takes place in the vapour form and such vapour movement has some practical implications in supplying water to drought resistant plants.

Entry of Water into Soil

Infiltration: Infiltration refers to the downward entry or movement of water into the soil surface

- It is a surface characteristic and hence primarily influenced by the condition of the surface soil.

- Soil surface with vegetative cover has more infiltration rate than bare soil

- Warm soils absorb more water than colder ones

- Coarse surface texture, granular structure and high organic matter content in surface soil, all help to increase infiltration

- Infiltration rate is comparatively lower in wet soils than dry soils

Factors affecting infiltration
    i. Clay minerals
    ii. Soil Texture
    iii. Soil structure
    iv. Moisture content
    v. Vegetative cover
    vi. Topography

Percolation: The movement of water through a column of soil is called percolation. It is important for two reasons.
    i. This is the only source of recharge of ground water which can be used through wells for irrigation
    ii. Percolating waters carry plant nutrients down and often out of reach of plant roots (leaching)

- In dry region it is negligible and under high rainfall it is high

- Sandy soils have greater percolation than clayey soil

- Vegetation and high water table reduce the percolation loss

Permeability: It indicates the relative ease of movement of water with in the soil. The characteristics that determine how fast air and water move through the soil is known as permeability. The term hydraulic conductivity is also used which refers to the readiness with which a soil transmits fluids through it.

DRAINAGE

The frequency and duration of periods when the soil is free from saturation with water. It controls the soil cum water relationship and the supply of nutrients to the plants.

- ◉ Drainage class
- ◉ Very poorly drained
- ◉ Poorly drained
- ◉ Imperfect
- ◉ Moderately well
- ◉ Well
- ◉ Somewhat excessive
- ◉ Excessive

Hysterisis

The moisture content at different tensions during wetting of soil varies from the moisture content at same tensions during drying. This effect is called as hysterisis. This is due to the presence of capillary and non capillary pores. The moisture content is always low during sorption and high during desorption. Hystersis phenomenon exists in soil minerals as a consequence of shrinking and swelling. Shrinking and swelling affect pore size on a microbasis as well as on the basis of overall bulk density.

.So, hystersis phenomenon occurs due to factors like shape and size of soil pores and their interconnection with each other pore configuration, nature of soil colloids bulk density of soil and entrapped air. The most important factor affecting hystersis is the entrapment of air in the soil under rewetting condition. This clogs some pores and prevent effective contact between others.

METHODS OF DETERMINATION OF SOIL MOISTURE

Two general types of measurements relating to soil water are ordinarily used

    i. By some methods the moisture content is measured directly or indirectly

    ii. Techniques are used to determine the soil moisture potential (tension or suction)

Measuring soil moisture content in laboratory

1. Gravimetric method: This consists of obtaining a moist sample, drying it in an oven at 105°C until it losses no more weight and then determining the percentage of moisture. The gravimetric method is time consuming and involves laborious processes of sampling, weighing and drying in laboratory.

2. Electrical conductivity method: This method is based upon the changes in electrical conductivity with changes in soil moisture. Gypsum blocks inside of with

two electrodes at a definite distance are apart used in this method. These blocks require previous calibration for uniformity. The blocks are buried in the soil at desired depths and the conductivity across the electrodes measured with a modified Wheatstone bridge. These electrical measurements are affected by salt concentration in the soil solution and are not very helpful in soils with high salt contents.

Measuring soil moisture potential in situ (field)

Suction method or equilibrium tension method: Field tensiometers measure the tension with which water is held in the soils. They are used in determining the need for irrigation. The tensiometer is a porous cup attached to a glass tube, which is connected to a mercury monometer. The tube and cup are filled with water and cup inserted in the soil. The water flows through the porous cup into the soil until equilibrium is established. These tension readings in monometer, expressed in terms of cm or atmosphere, measures the tension or suction of the soil. If the soil is dry, water moves through the porous cup, setting up a negative tension (or greater is the suction). The tensiometers are more useful in sandy soils than in fine textured soils. Once the air gets entrapped in the tensiometer, the reliability of readings is questionable.

11

SOIL ORGANIC MATTER

INTRODUCTION

Why do we care so much about soil organic matter? Is it some vague favorable connection with organic farming? What does the term "organic" mean in a scientific sense? Fortunately, our instinct that soil organic matter is a good thing is well-founded. Under most circumstances, soils with more organic matter are superior to those with less. The reasons for this may not always be the same, but that only makes it all the more important to understand as many of the implications of soil organic matter content as possible.

Decomposition

The rate of decomposition of organic matter depends on the soil's temperature, moisture, aeration, pH and nutrient levels.

The warmer and wetter the climate, the faster the rate of organic matter breakdown. Cooler areas have higher levels of soil organic matter because it does not break down as quickly in low temperatures.

Waterlogged organic matter breaks down very slowly because microorganisms necessary for decomposition cannot exist where there is no oxygen. Soils formed from waterlogged organic matter are known as *peats*, and contain a high percentage of organic matter.

Acid soils with low pH usually contain greater quantities of organic matter because microorganisms become less active as soil acidity increases.

BENEFITS OF ORGANIC MATTER

Improve soil structure

As organic matter decays to humus, the humus molecules 'cement' particles of sand, silt, clay and organic matter into aggregates which will not break down in water. This cementing effect, together with the weaving and binding effect of roots and fungal strands in the decomposing organic matter, makes the soil aggregates stable in water.

Improves drainage

These larger, stable aggregates have larger spaces between them, allowing air and water to pass through the soil more easily.

Holds moisture

The aggregates are also very effective in holding moisture for use by plants. Humus molecules can absorb and hold large quantities of water for use by plant roots.

Provides nutrients

Organic matter is an important source of nitrogen, phosphorus and sulfur. These nutrients become available as the organic matter is decomposed by microorganisms. Because it takes time for this breakdown to occur, organic matter provides a slow release form of nutrients. If crops are continually removed from the soil, there is no organic matter for microbes to feed on and break down into nutrients, so fewer nutrients are available to plants.

Improves cation exchange capacity

Humus molecules are *colloids*, which are negatively charged structures with an enormous surface area. This means they can attract and hold huge quantities of positively charged nutrients such as calcium, magnesium and potassium until the plant needs them. Clays also have this capacity, but humus colloids have a much greater CEC than clays.

How to increase soil organic matter levels

- **Grow perennial pasture:** A period under perennial, grass-dominant pasture is an effective way of increasing organic matter in farm soils. Short-lived annual grasses are a source of dead roots; perennial grasses are a source of leaf matter. Even short periods (1–2 years) under pasture can improve soil structure, even though the actual increase in organic matter may be small.

- **Grow cereal crops:** Cereal crops leave significant amounts of organic matter in their dead roots and stubbles after harvest.

- **Grow green manure crops:** Green manure crops provide protective cover until they are ploughed into the soil. Initially they provide a large increase in organic matter levels, but they break down rapidly to give only a small increase in long-term organic matter levels; also, the ploughing operation can do more harm than the good done by the organic matter.

- **Spread manure:** Bulky organic manures will increase organic matter, but frequent and heavy applications are needed to produce significant changes.

- **Use organic fertilisers:** Organic fertilisers applied in large amounts can boost organic matter levels but are generally less cost-effective as supplies of nutrients than inorganic fertilisers. Applied in small quantities, they are unlikely to have a significant effect on organic matter levels.

- **Keep cultivation to a minimum:** Cultivation breaks down the stable aggregates, exposing humus in the aggregates to air and faster decomposition. Direct drill techniques allow you to sow seed while leaving stubble residues on top of the soil, and leaving aggregates intact.

- **Concentrate organic matter:** An alternative to increasing inputs is to make more effective use of what is already there. Retain all organic additions, whether roots, stubble or manure, close to the surface. The stability of soil structure is related to the concentration of organic matter at the surface, not the total quantity present in the soil.

PROBLEMS WITH INCORPORATION

Incorporation of organic matter can present some problems.

- It is difficult to incorporate large quantities by cultivation.

- Green manure crops break down quickly and provide only a small increase in soil organic matter levels. Ploughing hastens the breakdown of humus and may counteract the small benefit from the crop itself.

- If organic matter is incorporated when the soil is wet, the soil may compact so that there is not enough oxygen available for microroganisms to decompose the organic matter. This may affect crop growth and nitrogen supply.

- Chemicals released from organic matter may reduce the rate of plant growth for a short time or have a toxic effect on young seedlings.

- Incorporating straw can also lead to a temporary shortage of available nitrogen for the planted crop, as the microorganisms will draw on the limited nitrogen in the decomposing straw.

SOIL ORGANIC MATTER: MEANS TO IMPROVE ROOT AERATION

Reducing Duration of Waterlogging

Management practices that promote SOM reduce risk and duration of waterlogging by increasing rate of water infiltration, which increases the time soil can receive rain before ponding occurs and reduces time required to drain from saturation to field capacity. Among the many measurable soil water variables, infiltration is the most commonly assessed, and we note the need to better establish relationships between infiltration, time to ponding, and drainage. It is also important to note, as reviewed by for no-till, that management practices that promote SOM can have a neutral effect on water infiltration in some cases, despite positive effects in majority of cases.

Accelerated infiltration associated with SOM is attributable to several soil features. The redistribution of soil mass to larger aggregate size classes associated with SOM helps to explain an increase in total or macro- porosity (>0.3–0.4 mm), although this effect is not detectable in all cases. SOM also stabilizes aggregates, minimizing their dissolution into smaller, and pore-clogging size fractions that seal the soil surface against water infiltration. The most dramatic effects of SOM on infiltration can likely be traced to earthworms and/ or termites and their creation of wide, continuous, vertically-oriented pores through which water flows preferentially. More abundant - or more active - soil fauna may be due in part to reduced disturbance associated with some SOM-promoting practices, e.g., no-till. However, close relationships between SOM and earthworm abundance without the confounding effect of disturbance also indicate a role for SOM as faunal substrate supply.

Few studies have attempted to link SOM-induced reductions in waterlogging with crop yield. , however, find crop yields increased from reduced surface soil waterlogging associated with no-till. The ability of SOM-mediated reductions in waterlogging to benefit crops are most likely when (1) crop is sensitive to waterlogging and (2) rainfall intensity can be mediated by SOM on a timescale relevant to waterlogging stress (neither drizzle nor deluge); and (3) soil is otherwise poorly-drained.

Promoting Aeration in Non-saturated Soils

If crops experience inadequate aeration in non-saturated soils, it is reasonable to expect that SOM would improve gas diffusivity given its effects on related parameters of soil structure. Few studies investigate the isolated effect of SOM on gas diffusivity, however, researchers find a positive relationship between SOM and gas diffusivity at field capacity across a soil texture gradient. Future work should examine net effects of SOM on O_2 diffusivity and consumption in soils.

COMPACTION

Soil Compaction Constrains Root Development

Soil compaction reduces crop yields and is quantified via either bulk density or mechanical impedance. MI estimates the force encountered by the elongation of a living root, and is consequential for crops because greater MI inflates the photosynthate required for root elongation. Although MI measurements ignore biopores used preferentially by roots, MI is more descriptive than bulk density because it is sensitive to soil water. Drying soils present increasing MI, and to isolate effects of water stress from compaction stress per se on crop development, researchers use experimental compaction.

Compaction studies indicate that reduced crop yield from compaction is due in large part to constraints on root development. A root restricted by soil compaction is generally thicker than a root in non-compacted soil, likely due to greater axial force needed to overcome compaction. Reductions in total number of roots, rate of root elongation, total root length, or root biomass are also reported. Root length is generally more reduced than root dry mass, indicating the accumulation of belowground photosynthate without commensurate expansion of soil-contacting surface area available for nutrient and water uptake. Compaction is sometimes characterized by a hardpan around 20 cm depth, which leads to restricted root access to subsoil and concentrated root development in the topsoil. This pattern of root development prevents crop access of deep soil water most implicated in crop drought resistance.

A single threshold MI for crop sensitivity is unlikely to serve universally, and not only because cultivars and crops differ in MI tolerance. Threshold MI values also likely depend on definition by energy required to extend roots or by crop yield penalty. The MI required to reduce root growth efficiency is likely less than required to affect yields, and yield penalty due to restricted roots can be counteracted somewhat by fertilization. Whatever the threshold, the detrimental effects of compaction on crops has generated attention toward means to reduce it.

Soil Organic Matter Reduces Compaction and is Associated with Root Channels

Reduced Compaction: More Water Transpired before Mechanical Impedance Limits Growth

Although SOM is often promoted for its ability to alleviate soil compaction and associated increases in MI, the generation of data confirming a negative SOM-MI relationship has been hampered by the convention of measuring MI in soils near field capacity. Even in large datasets, no relationship between SOM and MI in soil near field capacity is found, likely because very wet soils (<1–10 kPa) offer minimal MI regardless of SOM. It is as MI increases in drying soils that an effect of SOM becomes apparent.

We highlight two studies showing the effect of SOM in reducing MI as soils dry. and investigated soils differing only in SOM concentrations. With few exceptions, MI increased as soils approached permanent wilting point, but the increase in MI was not as much in higher SOM soils. In other words, SOM allows soil to become drier before reaching a potentially root-constraining MI. For instance, find an MI of 1.5 MPa is reached in the 1% OM soil at <-100 kPa, whereas the 3% OM soil reaches the same MI at about <-200 kPa. For these soils, the difference in volumetric water content between -100 and -200 kPa is <0.02 m^{-3} H_2O m^{-3} soil, or <5 mm of water when considered over the top 25 cm. While added organic amendments to glacial till to create fixed SOM percentages, their results resemble those of, who compared fallow to grassland soils. The extent to which management-induced SOM benefits crops via reductions in compaction likely vary with context, particularly those relevant to MI thresholds (see section "Soil compaction constrains root development").

Root Channels to Subsoil Water

Although not connected to the physical or biological properties of SOM, management practices that promote SOM may also alleviate the effects of compaction by facilitating crop root access to the subsoil. Deep-rooted cover crops or perennial crops create root channels to subsoil, which are used by subsequent cash crops (;). Crops are most likely to benefit from these root channels if subsoil is compacted and if crops experience sufficient water stress for subsoil water stores to be relevant.

Where Does Soil Organic Matter Come From?

As indicated on the previous page, most soil organic matter, at one time, was alive. Plants and animals that live in the soil also die in the soil and contribute their remains to the "pool" of materials. If it all stopped there, though, we would be hip-deep or worse in plant and animal parts. When the plant or animal dies, smaller organisms begin their work. Insects, including earthworms, are primary decomposers because they begin the work of breaking down the remains. Fungi, such as those you find on the forgotten leftovers in your refrigerator, soon follow. Most of the work, however, is done by the smallest soil organisms, the bacteria. Bacteria have two self-serving functions in working on organic remains. First, they use some of the carbon for energy, much as we do when we eat a sandwich. Second, they assimilate some of the carbon (and other elements in the remains) for their bodies.

When bacteria use carbon for energy, they usually produce carbon dioxide gas. This gas seeps out of the soil and is lost as a possible source of soil organic matter. The carbon that bacteria use for their bodies, however, is solid. Through many generations of reproduction and death of bacteria, this carbon begins to look entirely different from the original plant or animal remains. It has become humus. Even though humus is organic, it is more complex than the plant or animal from which it came. It is not accurate to say that humus is simply decomposed organic matter. One result of the complexity of humus is that it becomes more difficult for other microorganisms to break down. For this reason, humus can have a lifetime of tens of years in a soil.

The previous explanation is what occurs when the process works as it should. If bacteria are absent or inactive for any reason, the process slows or stops. By far the most common reason for bacteria to be inactive is if the soil is cold. There is little organic matter decomposition and humus formation in the winter or in permanently cold climates. Vivid examples of this are the well-preserved bodies of prehistoric-age peoples recently found in cold mountain regions. Other reasons for slowed decomposition are an absence of oxygen and water. These two points are really the opposite of each other because the most common cause for a lack of oxygen in soil is if the soil is flooded. Swamps and peat bogs have large amounts of poorly decomposed plant and animal remains because the water there excludes the oxygen which most bacteria need. The opposite extreme, desert climates, has so little water that bacteria also function poorly. This, possibly as much as the Egyptians' skills as embalmers, has helped preserve the mummies for thousands of years.

What is Good About Soil Organic Matter?

Soil organic matter does have many useful functions in soil. Probably the most important is its role in forming and maintaining good soil structure. Simply put, soil structure is just the way the soil particles are put together. Individual soil particles of the clays, silts, and even sands, are quite small. In a soil where these particles remain separate and exist as individual grains of sand, silt or clay, the soil can be hard, resistant to root growth, too wet in the spring because water does not drain away rapidly, and too dry in the summer because, when rains do come, water fails to soak in and ends up running off.

If the individual particles are "glued" together by humus or other cementing agents to form aggregates, a better soil results. Water can enter rapidly through the large pores between aggregates, but is held for plants to use in the small pores within aggregates. Roots can penetrate through an aggregated soil more easily to find water and nutrients.

What Kind of Harm Can Come From Soil Organic Matter?

Just because organic matter is so good doesn't mean that more is always better. Some of the problems with excessive soil organic matter are obvious. For example, residue from a corn crop can choke planting equipment for a following crop. There are more possible problems with soil organic matter that aren't so obvious, though. Certain plants produce materials that are harmful to other plants. Alternatively, some organic material is released as a plant decomposes that suppresses the growth of other plants. Both of these effects are called allelopathy. Some examples of allelopathy include: materials leached from the bark of black walnut trees suppressing growth of other plants around the base of the tree; Johnsongrass reducing yields in an infested field because of materials released from the roots; and wheat residue reducing yields of a following sorghum crop.

An even more general problem is that many plant residues will actually remove available nutrients from the soil as they decompose. This is illustrated in Fig. 1 for nitrogen. Nitrogen is often the biggest plant nutrient problem because it is used in large quantities by both desirable plants and other organisms. Plant-available nitrogen is often low under the best of conditions.

When a plant residue is incorporated into soil, decomposition begins by the processes described above. At some point, bacteria go into action; and, in using the plant residue for food, these bacteria require more nitrogen than the plant residue can supply. This can be seen most easily by examining the carbon-nitrogen ratio (C/N). This ratio is simply the ratio of the weight of carbon to the weight of nitrogen in a plant material, soil, or any organism. As an example, C/N for sawdust is about 500/1. This means sawdust contains about 500 pounds of carbon for every one pound of nitrogen. In contrast, wheat straw has a C/N of about 80/1, rotten manure about 20/1, and young alfalfa hay about 13/1. The organic matter in the surface soil which we would typically find in Nebraska has a C/N of about 11/1. Bacteria, on the other hand, have C:N values of about 5/1, so they need much more nitrogen than they get from the plant materials they "eat". What can they do?

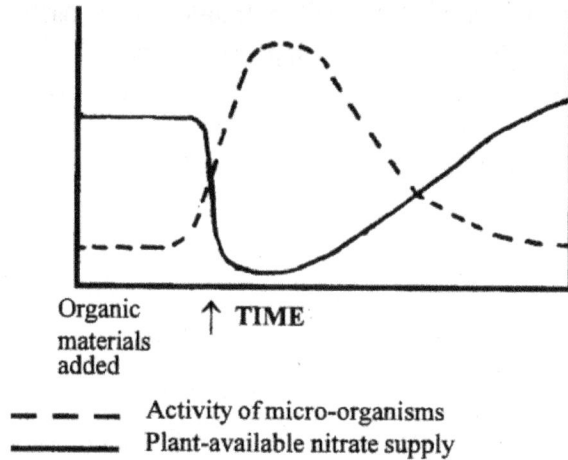

Fig. 1. An example of temporary nitrogen soil loss due to the incorporation of high C/N ratio residue into the soil.

One alternative is simply to slow down and not decompose all of the plant residue at once. This is one reason why some plant materials decompose more slowly than others. The bacteria don't have the nitrogen they need to decompose the residue quickly. Cornstalks (C/N of about 60/1) and wood chips can remain in a soil for months or years for this reason. Besides simply slowing down, however, the bacteria act as very effective scavengers. They "mine" the soil for every bit of available nitrogen that can be found. Most of this available nitrogen is in the form of nitrate and ammonium. The problem arises when the crop being grown needs available nitrogen at the same time. Bacteria are better competitors than crops for the limited nitrogen resource, so the crops do without until the bacteria die and release the nitrogen they have stored inside themselves back into the soil.

This means that most plant materials, but especially those with a high C/N ratio, should not be incorporated into a soil near the time when a desirable plant will also need nitrogen. If incorporated, a temporary nitrogen deficiency will result in the plant. After all, nitrogen was added to the soil when the residue was added in the first place; but it will take weeks or months, depending on the C/N of the residue and some other factors, for the soil to return to a state where the plant will have adequate supplies of available nitrogen. In this time, plant growth may have been so depressed that yield reductions will be substantial.

What can we do to remedy the situation? One obvious answer would be not to put plant residues into the soil. This would not usually be the best answer because of the many positive effects of returning plant residues to the soil. An alternative is to avoid putting plant residues in the soil when plant nitrogen demand is likely to be high. This could involve simply moving time of incorporation to the fall after harvest, rather than in spring immediately before planting. Applying additional nitrogen prior to incorporation would speed up the decomposition of the organic matter. On a smaller scale, "pre-decomposing" the plant residue by composting helps reduce or eliminate the temporary nitrogen deficiency that arises when high C/N materials are incorporated. Composting gives the bacteria time to work on the plant residue and lowers its C/N ratio before the compost is incorporated. This option is probably most practical for a homeowner dealing with yard waste, such as leaves or grass clippings. A good target value for C/N ratio is the same as the soil, 11/1, although incorporating plant material with C/N up to 20/1 will not usually have a serious or long-term effect on available soil nitrogen.

Why Does Soil Organic Matter Content Differ From Soil to Soil and Change?

Very few soils started out with substantial amounts of organic matter. This is mainly because most soil parent materials are geologic in origin. That is, they are developed from rock, or at least have been moved some distance before coming to rest to form a "new" soil. Because it takes plant growth to develop most soil organic matter, very little organic matter survives the temperatures, pressures and disturbance involved with geologic processes. Once a parent material comes to rest and plants begin to grow, however, organic matter starts to accumulate.

The two major natural variables which affect how much organic matter accumulates in a soil are **temperature** and *moisture*. Temperature affects organic matter accumulation in two ways. First, plants tend to grow faster and produce more total mass as temperatures increase. Everybody's mental image of a tropical rain forest involves lush, thick vegetation. Secondly, however, and usually overcoming the first point is that, as temperatures increase, microbial activity, including the activity of decomposing microorganisms, also increases. Going back to our mental picture of a tropical rain forest, it is surprising to find relatively little organic matter persisting in the soils of the forest floor. Microbial activity is just too intense to allow organic matter to accumulate.

Moisture, on the other hand, is a little more predictable in its effect on soil organic matter. As could be guessed, as rainfall increases, total plant production of organic matter increases, and so soil organic matter increases. Things get a little tricky, however, when the temperature and moisture effects are put together.

The west-to-east trend is because of increasing rainfall, while the south-to-north trend is because of lower temperatures preserving the organic matter that is produced. These trends hold up even into the Canadian prairies, despite their lower rainfall.

There are, of course, many natural and human-induced exceptions to these trends. One natural exception occurs where water accumulates to a degree that the soil is flooded for long periods of the year. In swamps or bogs, the excess water produces a shortage of oxygen which the decomposing bacteria need for their work. As a consequence, organic matter builds up regardless of the temperature, until the swamp is drained by natural or human causes. These locations are the source of peat available in garden stores.

One major human-induced factor which can change soil organic matter content is cropping. In Nebraska, the organic matter content of our prairie soils before the coming of white settlers was as high as six percent. Cropping, which involved a change from perennial grasses to annual plants, removed at least some of the organic matter produced in harvest and stirred the soil during tillage. Tillage opens up more of the soil to oxygen and increases microorganism breakdown of organic matter, at least temporarily. Within just a few years, organic matter content of a tilled soil can decrease to half of what it was in its previous prairie state. Managing a cropped soil with less disturbance, mainly by reducing tillage, will allow the organic matter content to rise somewhat, but probably not to the level of a native grassland.

Finally, plant growth and land use change not only the amount of total organic matter in a soil but the way in which that organic matter is distributed with depth.

The forest soil has high organic matter content at the surface, where leaf drop has contributed. Tree roots die off very little from season to season, however, so organic matter decreases rapidly with depth. In the native grassland, organic matter content is also high at the surface, but drops off slowly with depth because the fine roots die off year by year, and contribute organic matter to a greater depth than in a forest. In a native grassland soil which has been plowed, the organic matter content would be lower overall, and mixing from tillage would level the content of the top few inches.

SECONDARY EFFECTS OF SOIL ORGANIC MATTER

The previous discussion covered some of the direct effects of soil organic matter, mainly on soil structure. There are other effects of soil organic matter which are not so visible, but which may be just as important as improved structure.

Organic matter is a reservoir of certain plant nutrients, especially nitrogen, phosphorus, and sulfur. Many soils have 90 percent or more of their total nitrogen present in organic materials. The figures are smaller for phosphorus and sulfur, but are still significant. These organic reserves act as a bank account of nitrogen, phosphorus, and sulfur for plants. Actually, a better analogy might be that organic materials are a nutrient variable rate annuity. The plant can't use organic nitrogen any time it wants, just like you can't access a variable rate annuity until it comes to term. Organic forms of nutrients must be mineralized, or converted to inorganic forms, to be useful to plants. This is largely carried out by microorganisms as a byproduct of their use of organic matter for themselves.

A second indirect benefit of organic matter in soils is in increasing a quantity called cation exchange capacity (CEC). Cation exchange capacity is the ability of a soil to hold certain elements which have a positive charge, called cations. Some plant nutrients, such as potassium and calcium are cations, so a soil with a higher CEC will generally be more fertile because of its higher content of these nutrients. While there are exceptions to this generalization, it provides one more reason to see higher organic matter content as a benefit to the soil and to the crop being grown on the soil.

12

SOIL THERMAL PROPERTIES

The primary thermal properties of soil, or any substance, are the heat capacity and the thermal conductivity. The heat capacity can be defined per unit mass, in which case it is often called the specific heat, or per unit volume, in which case it is called the volumetric heat capacity. Sometimes it is useful to consider the ratio of the thermal conductivity to the volumetric heat capacity, and this ratio is called the thermal diffusivity. We will define and consider each of these in turn below. Knowledge of the soil thermal properties is necessary to predict how soil temperatures vary in space and time. Sensors which measure soil thermal properties can be used monitor soil water content nondestructively. Soil thermal properties also play a role in several remote-sensing based approaches for estimating soil moisture across large regions.

The thermal properties of soils are a component of soil physics that has found importants uses in engineering, climatology and agriculture. These properties influence how energy is partitioned in the soil profile. While related to soil temperature, it is more accurately associated with the transfer of heat throughout the soil, by radiation, conduction and convection.

MAIN SOIL THERMAL PROPERTIES

Volumetric heat capacity, SI units: $Jm-3K^{-1}$

Thermal conductivity, SI units: $W.m-1K^{-1}$

Thermal diffusivity, SI units: $m^2.s$

THERMAL CONDUCTIVITY

The soil ***thermal conductivity*** (λ) is the ratio of the magnitude of the conductive heat flux through the soil to the magnitude of the temperature gradient ($W\ m^{-1}\ °C^{-1}$). It is a measure of the soil's ability to conduct heat, just as the hydraulic conductivity is a measure of the soil's ability to "conduct" water. Soil thermal conductivity is influenced by a wide range of soil characteristics including:

- ◉ Air-filled porosity

- ◉ Water content

- ◉ Bulk density

- ◉ Texture

- ◉ Mineralogy

- ◉ Organic matter content

- ◉ Soil structure

- ◉ Soil temperature

Among common soil constituents, quartz has by far the highest thermal conductivity and air has by far the lowest thermal conductivity (Table 1). Often, the majority of the sand-sized fraction in soils is composed primarily of quartz, thus sandy soils have higher thermal conductivity values than other soils, all other things being equal. Since the thermal conductivity of air is so low, air-filled porosity exerts a dominant influence on soil thermal conductivity. Soil thermal conductivity increases as water content increases, but not in a purely linear fashion. For dry soil, relatively small increases in the water content can substantially increase the thermal contact between mineral particles because the water adheres to the particles, resulting in a relatively large increase in the thermal conductivity.

Table 1. Thermal conductivity, density, and specific heat of common soil constituents at 10 °C

Soil constituent	Thermal conductivity	Density	Specific heat
	$W\ m^{-1}\ °C^{-1}$	$g\ cm^{-3}$	$J\ g^{-1}\ °C^{-1}$
Quartz	8.8	2.66	0.75
Clay minerals	3	2.65	0.76
Soil organic matter	0.3	1.3	1.9
Water	0.57	1.00	4.18
Ice (0 °C)	2.2	0.92	2.0
Air	0.025	0.00125	1.0

HEAT CAPACITY

Soil *volumetric heat capacity* (*C*) is the amount of energy required to raise the temperature of a unit volume of soil by one degree ($J\ m^{-3}\ °C^{-1}$). Unlike thermal conductivity, volumetric heat capacity increases strictly linearly as soil water content increases. Volumetric heat capacity is also a linear function of bulk density. The volumetric heat capacity can be calculated by

$$C = \rho_b c_s + \rho_w c_w \theta \qquad (1)$$

where ρ_b is the soil bulk density (g cm^{-3}), cs is the specific heat of the soil solids (J g^{-1} °C^{-1}), ρ_w is the density of water (g cm^{-3}), c_w is the specific heat of water, and θ is the volumetric water content (cm^3 cm^{-3}). To increase the temperature of wetter, denser soil requires more energy than to increase the temperature of drier, less dense soil, which has a lower volumetric heat capacity. This is one factor that can contribute to lower soil temperatures and delayed crop development in soils managed with no tillage.

THERMAL DIFFUSIVITY

The soil thermal diffusivity is the ratio of the thermal conductivity to the volumetric heat capacity (m^2 s^{-1}). It is an indicator of the rate of at which a temperature change will be transmitted through the soil by conduction. When the thermal diffusivity is high, temperature changes are transmitted rapidly through the soil. Logically, soil thermal diffusivity is influenced by all the factors which influence thermal conductivity and heat capacity. Thermal diffusivity is somewhat less sensitive to soil water content than are thermal conductivity and volumetric heat capacity. The thermal diffusivity is a particularly useful parameter to aid in understanding and modeling soil temperatures, which is the next topic we will consider.

SOIL TEMPERATURE – SOIL AIR – GASEOUS EXCHANGE

Soil Temperature

Soil temperature is an important plant growth factor like air, water and nutrients. Soil temperature affects plant growth directly and also indirectly by influencing moisture, aeration, structure, microbial and enzyme activities, rate of organic matter decomposition, nutrient availability and other soil chemical reactions. Specific crops are adapted to specific soil temperatures. Apple grows well when the soil temperature is about 18°C, maize 25°C, potato 16 to 21°C, and so on.

Sources of soil heat

The sources of heat for soil are solar radiation (external), heat released during microbial decomposition of organic matter and respiration by soil organisms including plants and the internal source of heat is the interior of the Earth - which is negligible. The rate of solar radiation reaching the earth's atmosphere is called as solar constant and has a value of 2 cal cm^{-2} min^{-1}. Major part of this energy is absorbed in the atmosphere, absorbed by plants and also scattered. Only a small part of it reaches soil. Thermal energy is transmitted in the form of thermal infrared radiation from the sun across the space and through the atmosphere.

Factors affecting soil temperature

The average annual soil temperature is about 1°C higher than mean annual air temperature. Soil temperature is influenced by climatic conditions. The factors that affect the transfer of heat through the atmosphere from sun affect the soil temperature also.

Environmental factors

Solar radiation

The amount of heat received from sun on Earth's surface is 2 cal cm^{-2} min^{-1}. But the amount of heat transmitted into soil is much lower. The heat transmission into soil depends on the angle on incident radiation, latitude, season, time of the day, steepness and direction of slope and altitude. The insulation by air, water vapour, clouds, dust, smog, snow, plant cover, mulch etc., reduces the amount of heat transferred into soil.

Soil factors

a. **Thermal (Heat) capacity of soil:** The amount of energy required to raise the temperature by 1°C is called heat capacity. When it is expressed per unit mass (Calories per gram), then it is called as specific heat. The specific heat of water is 1.00 cal g^{-1} where the specific heat of a dry soil is 0.2 cal g^{-1}. Increasing water content in soil increases the specific heat of the soil and hence a dry soil heats up quickly than a moist soil.

b. **Heat of vaporization:** The evaporation of water from soil requires a large amount of energy, 540 kilocalories kg-1 soil. Soil water utilizes the energy from solar radiation to evaporate and thereby rendering it unavailable for heating up of soil. Also the thermal energy from soil is utilized for the evaporation of water, thereby reducing the soil temperature. This is the reason that surface soil temperatures will be sometimes 1 to 6°C lower than the sub-surface soil temperature. That is why the specific heat of a wet soil is higher than dry soil.

c. **Thermal conductivity and diffusivity:** This refers to the movement of heat in soils. In soil, heat is transmitted through conduction. Heat passes from soil to water about 150 times faster than soil to air. So the movement of heat will be more in wet soil than in dry soil where the pores will be occupied with air. Thermal conductivity of soil forming materials is 0.005 thermal conductivity units, and that of air is 0.00005 units, water 0.001 units. A dry and loosely packed soil will conduct heat slower than a compact soil and wet soil.

d. **Biological activity:** Respiration by soil animals, microbes and plant roots evolve heat. More the biological activity more will be the soil temperature.

e. **Radiation from soil:** Radiation from high temperature bodies (Sun) is in short waves (0.3 to 2.2°) and that from low temperature bodies (soil) is in long waves (6.8 to 100°) Longer wavelengths have little ability to penetrate water vapour, air and glass and hence soil remains warm during night hours, cloudy days and in glass houses.

f. **Soil colour:** Colour is produced due to reflection of radiation of specific wavelengths. Dark coloured soils radiate less heat than bright coloured soils. The ratio between the incoming (incident energy) and outgoing (reflected energy) radiation is called albedo. The larger the albido, the cooler is the soil. Rough surfaced soil absorbs more solar radiation than smooth surface soils.

$$\text{Albido} = \frac{\text{Reflected energy}}{\text{Incidentenergy}}$$

g. **Soil structure, texture and moisture:** Compact soils have higher thermal conductivity than loose soils. Natural structures have high conductivity than disturbed soil structures. Mineral soils have higher conductivity than organic soils. Moist soil will have uniform temperature over depth because of its good conductivity than dry soils.

h. **Soluble salts:** Indirectly affects soil temperature by influencing the biological activities, evaporation etc.

Soil Air

Soil air is a continuation of the atmospheric air. Unlike the other components, it is constant state of motion from the soil pores into the atmosphere and from the atmosphere into the pore space. This constant movement or circulation of air in the soil mass resulting in the renewal of its component gases is known as soil aeration.

Composition of Soil Air

The soil air contains a number of gases of which nitrogen, oxygen, carbon dioxide and water vapour are the most important. Soil air constantly moves from the soil pores into the atmosphere and from the atmosphere into the pore space. Soil air and atmospheric air differ in the compositions. Soil air contains a much greater proportion of carbon dioxide and a lesser amount of oxygen than atmospheric air. At the same time, soil air contains a far great amount of water vapour than atmospheric air. The amount of nitrogen in soil air is almost the same as in the atmosphere.

Composition of soil and atmospheric air

	Percentage by volume		
	Nitrogen	Oxygen	Carbon dioxide
Soil air	79.2	20.60	0.30
Atmospheric air	79.9	20.97	0.03

Factors Affecting the Composition of Soil Air

1. **Nature and condition of soil:** The quantity of oxygen in soil air is less than that in atmospheric air. The amount of oxygen also depends upon the soil depth. The oxygen content of the air in lower layer is usually less than that of the surface soil. This is possibly due to more readily diffusion of the oxygen from the atmosphere into the surface soil than in the subsoil. Light texture soil or sandy soil contains much higher percentage than heavy soil. The concentration of CO_2 is usually greater in subsoil probably due to more sluggish aeration in lower layer than in the surface soil.

2. **Type of crop:** Plant roots require oxygen, which they take from the soil air and deplete the concentration of oxygen in the soil air. Soils on which crops are grown

contain more CO_2 than fallow lands. The amount of CO_2 is usually much greater near the roots of plants than further away. It may be due to respiration by roots.

3. **Microbial activity:** The microorganisms in soil require oxygen for respiration and they take it from the soil air and thus deplete its concentration in the soil air. Decomposition of organic matter produces CO_2 because of increased microbial activity. Hence, soils rich in organic matter contain higher percentage of CO_2.

4. **Seasonal variation:** The quantity of oxygen is usually higher in dry season than during the monsoon. Because soils are normally drier during the summer months, opportunity for gaseous exchange is greater during this period. This results in relatively high O_2 and low CO_2 levels.

Temperature also influences the CO_2 content in the soil air. High temperature during summer season encourages microorganism activity which results in higher production of CO_2.

Exchange of Gases between Soil and Atmosphere

The exchange of gases between the soil and the atmosphere is facilitated by two mechanisms

1. **Mass flow:** With every rain or irrigation, a part of the soil air moves out into the atmosphere as it is displaced by the incoming water. As and when moisture is lost by evaporation and transpiration, the atmospheric air enters the soil pores. The variations in soil temperature cause changes in the temperature of soil air. As the soil air gets heated during the day, it expands and the expanded air moves out into the atmosphere. On the other hand, when the soil begins to cool, the soil air contracts and the atmospheric air is drawn in.

2. **Diffusion:** Most of the gaseous interchange in soils occurs by diffusion. Atmospheric and soil air contains a number of gases such as nitrogen, oxygen, carbon dioxide etc., each of which exerts its own partial pressure in proportion to its concentration.

The movement of each gas is regulated by the partial pressure under which it exists. If the partial pressure on one of the gases (i.e. carbon dioxide) is greater in the soil air than in the atmospheric air, it (CO_2) moves out into the atmosphere. Hence, the concentration of CO_2 is more in soil air.

On the other hand, partial pressure of oxygen is low in the soil air, as oxygen present in soil air is consumed as a result of biological activities. The oxygen present in the atmospheric air (partial pressure of O_2 is greater) therefore, diffuses into the soil air till equilibrium is established. Thus, diffusion allows extensive movement and continual change of gases between the soil air and the atmospheric air. Oxygen and carbon dioxide are the two important gases that take in diffusion.

Importance of Soil Aeration

1. **Plant and root growth:** Soil aeration is an important factor in the normal growth of plants. The supply of oxygen to roots in adequate quantities and the removal of CO_2 from the soil atmosphere are very essential for healthy plant growth. When

the supply of oxygen is inadequate, the plant growth either retards or ceases completely as the accumulated CO2 hampers the growth of plant roots. The abnormal effect of insufficient aeration on root development is most noticeable on the root crops. Abnormally shaped roots of these plants are common on the compact and poorly aerated soils. The penetration and development of root are poor. Such undeveloped root system cannot absorb sufficient moisture and nutrients from the soil

2. **Microorganism population and activity:** The microorganisms living in the soil also require oxygen for respiration and metabolism. Some of the important microbial activities such as the decomposition of organic matter, nitrification, Sulphur oxidation etc depend upon oxygen present in the soil air. The deficiency of air (oxygen) in soil slows down the rate of microbial activity. For example, the decomposition of organic matter is retarded and nitrification arrested. The microorganism population is also drastically affected by poor aeration.

3. **Formation of toxic material:** Poor aeration results in the development of toxin and other injurious substances such as ferrous oxide, H_2S gas, CO_2 gas etc in the soil.

4. **Water and nutrient absorption:** A deficiency of oxygen has been found to check the nutrient and water absorption by plants. The energy of respiration is utilized in absorption of water and nutrients. Under poor aeration condition (this condition may arise when soil is water logged), plants exhibit water and nutrient deficiency

5. **Development of plant diseases:** Insufficient aeration of the soil also lead to the development of diseases. For example, wilt of gram and dieback of citrus and peach.

INFLUENCE OF SOIL TEMPERATURE AND AIR ON PLANT GROWTH

Effect of soil temperature on plant growth

    a. Soil temperature requirements of plants: The soil temperature requirements of plants vary with the species. The temperature at which a plant thrives and produces best growth is called optimum range (temperature). The entire range of temperature under which a plant can grow including the optimum range is called growth range. The maximum and minimum temperatures beyond which the plant will die are called survival limits.

Range	Maize (°C)	Wheat (°C)
Optimum range	25 - 35	15 - 27
Growth range	Oct-39	May-35
Survival limits	0 - 43	0 - 43

    b. Availability of soil water and plant nutrients: The free energy of water increases with temperature. Up to wilting point limit, warming of soil increases water availability beyond which it decreases. Low temperatures reduce the nutrient availability, microbial activities and root growth and branching. The ability to absorb nutrients and water by plants reduces at low temperatures.

SOIL TEMPERATURE MANAGEMENT

Use of organic and synthetic mulches: Mulches keep soil cooler in hot summer and warm in cool winter.

Soil water management: High moisture content in humid temperate region lowers soil temperature.

Tillage management: Tilling soil to break the natural structure reduces the heat conductance and heat loss. A highly compact soil looses heat faster than loose friable soil.

Methods of measuring soil temperature: Mercury soil thermometers of different lengths, shapes and sizes with protective cover are buried at different depths to measure the temperature. Thermo couple and thermister based devices are also available. Infra-red thermo meters measure the surface soil temperature. Automatic continuous soil thermographs record the soil temperatures on a time scale. The International Meteorological Organization recommends standard depths to measure soil temperatures at 10, 20, 50 and 100 cm.

HEAT FLOW AND THERMAL EFFECTS IN SOILS

Heat Transport in Soils

There are several pathways for the transport of energy in the form of heat through soils. Heat may be transported in soils by conduction, radiation, and by convection with air or water flow (Fig. 1).

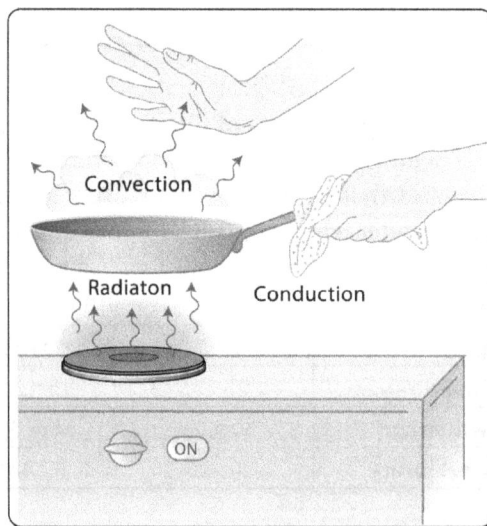

Fig 1 Cook stove illustration of the role of radiation, convection, and conduction in heat transfer.

We recall from our discussion of the energy balance that radiative transfer in the subsurface would be due to black body radiation (also called longwave exchange). Thinking back, you will remember that black-body exchange is proportional to the fourth power of the temperature difference between the two bodies.

Your toaster has glowing red wires that are at about 1300 °K, while the toast is at about 300 °K, so the difference is 900 °K, and the rate of energy exchange will be proportional to $(900 °K)^4$ times the emissivities of the toast and wire (not far from one, especially for whole wheat bread).

Now in the soil there might be 20 grains of sand per cm, and the temperature gradient might be 1°C per cm (in a highest case), so the temperature difference between adjacent grains will be something like 1/20 x 1°C = 0.05°C. So the rate of energy exchange between the grains will be $(0.05/900)^4 = 10^{-17}$ times smaller than the exchange your bread feels. As you may have surmised, radiative transfer is insignificant under these conditions.

You might be thinking that we will similarly dispense with convection, after all, how much heat can really be carried by the slow-moving water and air in soil? While in many cases, especially natural soils within a meter or two of the surface, your intuition is correct in figuring that convection generally plays a small role. However, the story is more interesting in cases of high flux (irrigated agriculture) or deep unsaturated zones.

Especially far from the surface, where there are very low thermal gradients to drive conduction, convection of percolating water can strongly effect the subsurface temperature. An extreme case of convection in the subsurface are springs, where water typically emerges at a temperature close to that of the regional aquifer temperature rather than the soil temperature. So next time you are either cooling yourself in a mountain stream in heat of the summer, or warming in a thermal hot spring in the winter, you can thank convection for your pleasure.

Another very interesting case of energy transport is via gas movement in the form of water vapor transport. In this case the energy transfer is in the form of "latent" heat, in that the energy will be consumed turning liquid water into water vapor, and released when the water vapor condenses. Because the latent heat is so large (2257 J/g) compared to the specific heat of water which is just XX J/°K-g. As we will discuss below, this latent heat transfer is typically driven by diffusive vapor movement, where water vapor moves from regions of high water vapor concentration to low concentration (which in damp soils means from warm areas to cooler areas). Technically we would refer to this as a diffusive flux, but it does give rise to a net flux of water vapor which carries energy through a soil profile, so it must be kept front and center as we consider heat flow processes.

Having touched on the lesser players, it is time to turn to the most important process for heat flow in soils. Conduction of heat through matter involves transfer of kinetic energy at the molecular level, where molecules in warmer regions vibrate rapidly resulting in collisions with, or excitation of, their colder "neighbors." The flow of heat in soil and the exchange of heat with the atmosphere determine a soil's thermal regime. This regime is characterized mainly by the soil temperature with its temporal and spatial variations. The soil thermal regime affects many biological, chemical and physical processes including above and below ground plant growth development, microbial activity, the kinetics and transformation of nutrients, frost heaving, and many other processes.

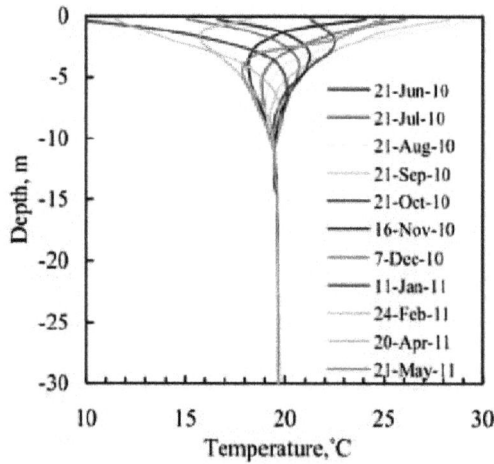

Fig. 2 Simulated seasonal temperature profiles. It is noteworthy that the same depth dependence is seen around the world, almost universally showing a loss of seasonal variation at about 10 m.

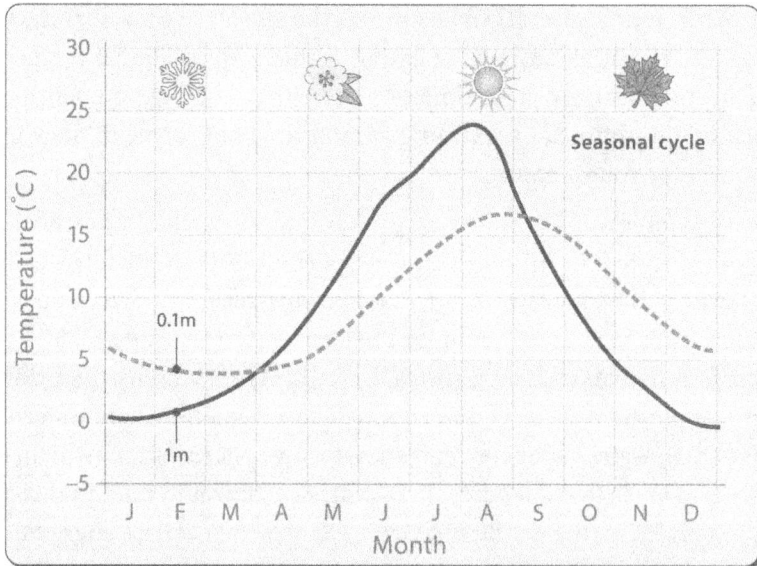

Fig. 3 Plots of the penetration of diurnal (a) and seasonal (b) thermal signals at various depths in a typical soil profile. Note that diurnal variations essentially lost by 50 cm depth, while seasonal variations in temperature penetrate to about 10 m, as seen in Figure 2.

Considering the soil-atmosphere interface, the heat capacity of the first 25 cm of soil buffers the air temperature, yielding warmer nights and cooler days (Fig. 3). This also works on a seasonal basis, where the top two meters of soils change temperature. Of course, from the nuclear reactions taking place in the center of the earth there is about 0.1 W/m² of thermal energy leaving the surface, which gives rise to an average °25C/km thermal gradient in the earth. This "ground heat flux" is truly essential to understanding the dynamics of local conditions and global climate.

Thermal heat conduction can also drive water movement near soils surfaces, especially in arid environment's. Fundamentally this reflects the fact that near the surface of the soil much of the water is lost in the form of water vapor, and the phase change from liquid to vapor forms required significant energy (think of how cold your hands feel after washing them when you are in dry air). To make this a bit more concrete, let's consider a morning when the atmosphere is at 15°C and 40% relative humidity (RH). The absolute humidity would be about 0.0075 gr H_2O/gr dry air. Let's suppose that the surface soil is dry, and that 5 cm below the surface the soil is moist, so has RH=100%, and has a temperature of 20°C.

Again, from the psychrometric chart, the absolute humidity 5 cm into the soil will be 0.015 gr H_2O/gr dry air. So we see there will be a gradient in humidity (0.15 gr H_2O/gr dry air/m), which will lead to a flux of water vapor toward the surface (Fick's law – we'll play with that later when we work on transport in soils). This will then lead to the lost water vapor to be replaced by evaporation of the soil water at the 5-cm depth, which will cool the soil locally. This energy may well be replaced by thermal conduction of heat stored from the previous day from deeper in the soil profile. These linked processes of vapor and thermal energy transport are subjects of much current research.

HEAT CONDUCTION IN SOIL

The most important features of heat transfer in soils may be cast in a form of heat conduction, which we will use as the first step in "building" the heat transfer equation. In as soil without mass movement (of water or gas), the vertical one dimensional flux density of heat J_H[W m⁻²]) in soil is described by *Fourier's law*:

$$J_H = \lambda \frac{dT}{dz} \qquad (2)$$

Relating the energy flux J_H to the soil thermal conductivity λ (J m⁻¹ s⁻¹°C⁻¹), and the gradient in temperature T (°C), with distance z, which could be the vertical spatial coordinate or soil depth.

While thermal conductivity in soils is in principle independent of water vapor movement, the λ in Eq. 2 should be considered as the *apparent* soil thermal conductivity, as latent heat transfer in the form of water vapor (that is energy liberation or consumption

Fig. 4 Joseph Fourier (1768 - 1830)

due to water changing between liquid and vapor states) cannot in practice be separated from conduction in moist soils; i.e.,

$$\lambda = \lambda^* + D_{vapor} * L,$$

Where λ^* is the instantaneous thermal conductivity, D_{vapor} is thermal vapor diffusivity, and L is latent heat of vaporization. We will discuss this more when we add mass movement to the equation.

A key observation is that ë depends on the minerals which make up the soil, the porosity of the soil, and the water content of the sample (Fig. 5). The thermal conductivity of the constituents of soil vary by 400-fold, with quartz having fifteen times the conductivity of water, which in turn is about twenty times as conductive as air (Table 2).

Table 2 Thermal conductivity of some of the typical constituents of soils

Constituent	[W m^{-1} °C^{-1}]
Quarts	8.8
Soil minerals (avg.)	2.9
Soil organic matter	0.25
Water (liquid)	0.57
Ice (at 0°C)	2.2
Air	0.025

Heath flow path

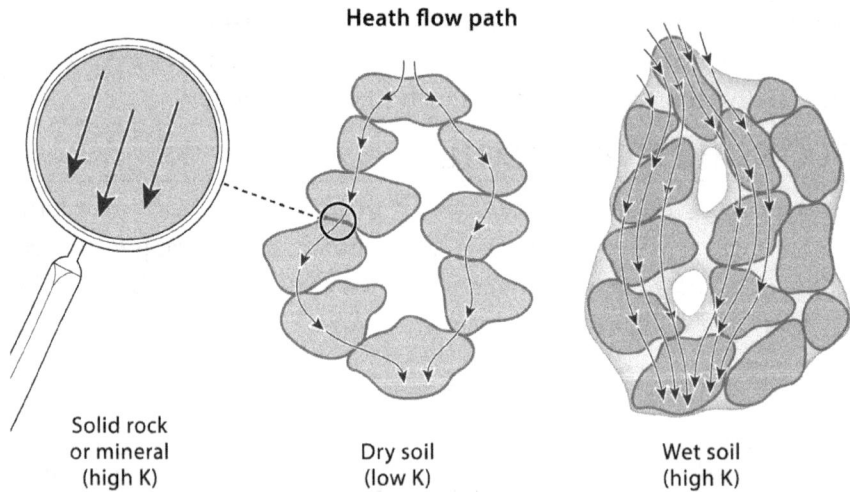

Solid rock or mineral (high K)   Dry soil (low K)   Wet soil (high K)

Fig. 5 Illustration of heat flow in soils, emphasizing the role of mineral composition, porosity, and water content.

Analogous to the weakest link controlling the strength of a chain, flow of heat is controlled largely by the contacts between the mineral particles. Water can act as bridge between grains, so there is often a great change in conductivity with the transition from completely dry to residual moisture content, facilitating the movement of water through the minerals. Conversely, the same amount of water added near saturation change thermal conductivity very little, since the minerals are carrying the bulk of the heat (Fig. 6).

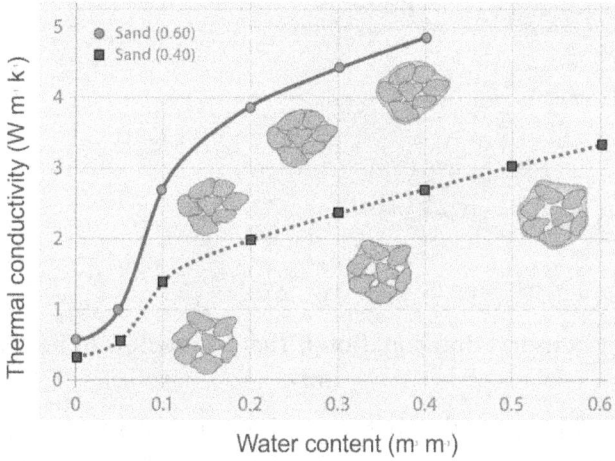

Fig. 6 Typical valued of thermal conductivity of sand and clay soils as a function of water content. Note the disproportionately large impact of the first 10% of volumetric water content which bridges between grains, providing the key pathways for heat flow.

Combining the heat flux with the equation for conservation of heat energy results in a general expression for conductive soil heat flow where soil temperature may vary in time and space (Figure 7a and b).

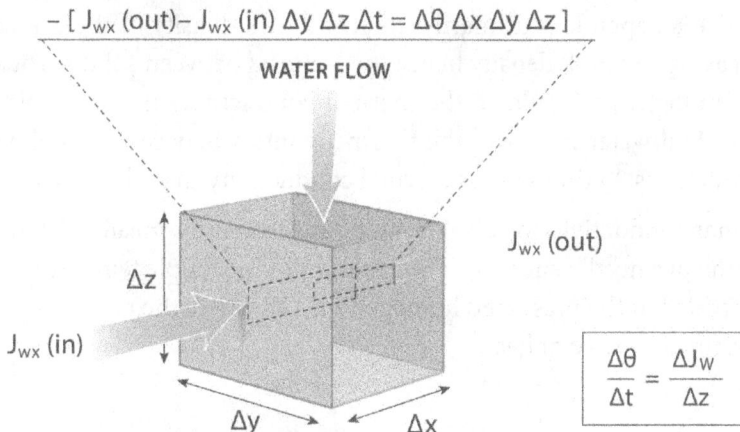

$$- [\, J_{wx} \, (\text{out}) - J_{wx} \, (\text{in}) \, \Delta y \, \Delta z \, \Delta t = \Delta \theta \, \Delta x \, \Delta y \, \Delta z \,]$$

WATER FLOW

J_{wx} (out)

Δz

J_{wx} (in)

$$\frac{\Delta \theta}{\Delta t} = \frac{\Delta J_w}{\Delta z}$$

Δy Δx

Fig. 7a. The conservation of energy follows exactly the same reasoning as the conservation of mass employed in deriving Richards equation.

Using the control volume approach as we did in the derivation of Richards Equation, we consider an infinitesimally small cube with sides of length Δx, Δy, Δz (Fig 7b). Let's take the center of the cube to be at point (x, y, z), so the sides of the cube will be above, below, in front, behind, left or right of this point adistances $\Delta x/2$, $\Delta y/2$, and $\Delta z/2$. The basic concept is that

Energy in - Energy out = change in energy   (3)

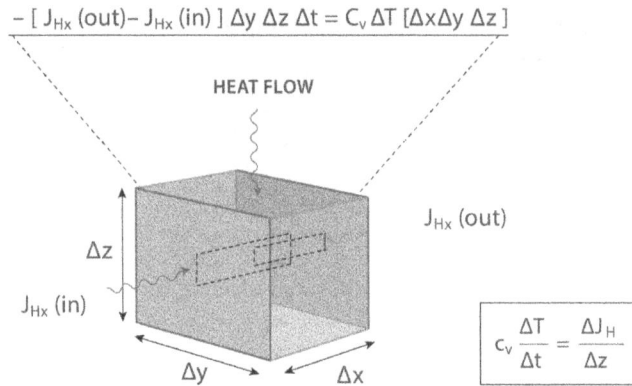

$$- [J_{Hx} (out)- J_{Hx} (in)] \, \Delta y \, \Delta z \, \Delta t = C_v \, \Delta T \, [\Delta x \Delta y \, \Delta z]$$

HEAT FLOW

J_{Hx} (out)

Δz

J_{Hx} (in)

Δy Δx

$$C_v \frac{\Delta T}{\Delta t} = \frac{\Delta J_H}{\Delta z}$$

Fig. 7b. The energy conservation equation if there were flow of heat in the x-direction only.

So if the energy flux vector (has both magnitude and direction) is given by J_H(x, y, z), and we will denote the component in the x-direction (this is a scalar) to be J_{Hx}(x, y, z). The overall plan is to estimate the flux at the surface of the cube as this central value adjusted by the gradient in flux times the distance the side of the cube is from the center (e.g., $\Delta x/2$).

Thermal Conductivity

Thermal conductivity is defined as the amount of heat transferred through a unit area in unit time (heat flux density) under a unit temperature gradient. The soil thermal conductivity (λ) is dependent primarily upon the bulk density of the soil and its water content. Increasing soil bulk density hence the contacts between solid particles increases the thermal conductivity. Similarly, the thermal conductivity increases with increasing water content . Soil water improves the thermal contact between the soil particles, and replaces air which has 20 times lower thermal conductivity than does water (Table 3).

The thermal conductivity of a soil may be measured by means of a transient-heat probe consisting of a needle encasing a heater and thermocouple temperature sensor. The sensor is embedded in the presumed homogenous soil where the needle approximates an infinitely long linear source of heat.

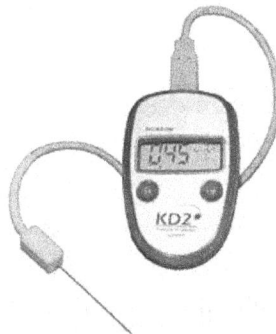

Fig. 8. A commercially available single-probe meter to measure thermal properties of media which allow the insertion of a needle

Table 3 Thermal Conductivities of Soil Constituents at 10°C

Constituent	λ [mcal (cm s °C)⁻¹]	λ [W m⁻¹ °C⁻¹]
Quartz	21	8.8
Soil minerals (avg.)	7	2.9
Soil organic matter	0.6	0.25
Water (liquid)	1.37	0.57
Ice (at 0°C)	5.2	2.2
Air	0.06	0.025

Heat is generated for a short time by application of a constant electric current along the needle, while measuring its temperature rise with respect to time using a thermocouple collocated with the heater.

Thermal Diffusivity

Thermal diffusivity is the ratio of the thermal conductivity to the volumetric heat capacity: $D_H = \lambda/c_v$. It may be estimated from measurements of λ and c_v, or it may be measured directly. Thermal diffusivity at first increases rapidly with increasing water content, then decreases at a slower rate. This behavior results from the fact that while heat capacity (c_v) increases linearly with water content, ë increases most rapidly at low water contents. It is important to note that D_H is not monotonic for many soils – so that the same value may be found for two different water contents. This is a key feature that must be kept in mind if one seeks to determine water content from soil thermal properties: any expression dependent primarily on D_H will not be useful to estimate water content.

One method for measuring D_H directly is based on placing a heat source having constant temperature (T_s) in contact with the surface of a soil column having constant cross sectional area and insulated sides, initially at ambient temperature (T_0).

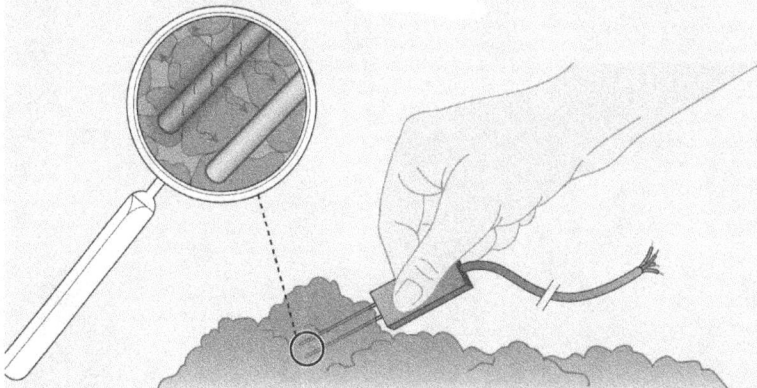

Fig. 9 The dual probe heat probe method allows both the measurement of the response within the needle which provides the heat, as well as a sensing needle which is separated by a known distance. Combining the two measurements constrains thermal properties much more than does the single probe approach.

DIURNAL AND SEASONAL VARIATIONS IN SOIL TEMPERATURE

In the field, soil thermal regime is characterized by periodic changes in response to the natural periodicity in atmospheric conditions controlling energy inputs to the soil surface. There is a diurnal (daily) cycle, as well as a superimposed seasonal cycle. These diurnal and seasonal cycles are perturbed by irregular meteorological events including cloudiness, warm and cold fronts, precipitation, etc.

We may identify three fundamental characteristics of diurnal and annual soil thermal regimes. First, because these energy inputs are cyclic, we observe diurnal and annual *temperature cycles* in response to these fluctuating inputs. Secondly, because much of the heat traveling down through a soil profile is utilized to change the local temperature of the soil, there is a decrease in energy flux at depth. Thus we observe the phenomenon of *amplitude damping*, or a reduction in the magnitude of these temperature cycles with increasing depth. Finally, because it takes time for heat to travel into and out of the soil, there is a delay in the time at which any specific location on the temperature cycle reaches a given point in the soil, and this *time lag* becomes more pronounced with increasing distance.

13

LAYER SILICATE CLAYS - GENESIS AND CLASSIFICATION

INTRODUCTION TO CLAY MINERALS & SOILS

Clay minerals are layer silicates that are formed usually as products of chemical weathering of other silicate minerals at the earth's surface. They are found most often in shales, the most common type of sedimentary rock. In cool, dry, or temperate climates, clay minerals are fairly stable and are an important component of soil. Clay minerals act as "chemical sponges" which hold water and dissolved plant nutrients weathered from other minerals. This results from the presence of unbalanced electrical charges on the surface of clay grains, such that some surfaces are positively charged (and thus attract negatively charged ions), while other surfaces are negatively charged (attract positively charged ions). Clay minerals also have the ability to attract water molecules. Because this attraction is a surface phenomenon, it is called adsorption (which is different from *absorption* because the ions and water are not attracted deep inside the clay grains). Clay minerals resemble the micas in chemical composition, except they are very fine grained, usually microscopic. Like the micas, clay minerals are shaped like flakes with irregular edges and one smooth side. There are many types of known clay minerals. Some of the more common types and their economic uses are described here:

Kaolinite

This clay mineral is the weathering product of feldspars. It has a white, powdery appearance. Kaolinite is named after a locality in China called Kaolin, which invented porcelain (known as china) using the local clay mineral. The ceramics industry uses it extensively. Because kaolinite is electrically balanced, its ability of adsorb ions is less than that of other clay minerals. Still, kaolinite was used as the main ingredient for the original formulation of the diarrhea remedy, Kaopectate.

Illite

Resembles muscovite in mineral composition, only finer-grained. It is the weathering product of feldspars and felsic silicates. It is named after the state of Illinois, and is the dominant clay mineral in midwestern soils.

Chlorite

This clay mineral is the weathering product of mafic silicates and is stable in cool, dry, or temperate climates. It occurs along with illite in midwestern soils. It is also found in some metamorphic rocks, such as chlorite schist.

Vermiculite

This clay mineral has the ability to adsorb water, but not repeatedly. It is used as a soil additive for retaining moisture in potted plants, and as a protective material for shipping packages.

Smectite

This clay mineral is the weathering product of mafic silicates, and is stable in arid, semi-arid, or temperate climates. It was formerly known as *montmorillonite*. Smectite has the ability to adsorb large amounts of water, forming a water-tight barrier. It is used extensively in the oil drilling industry, civil and environmental engineering (where it is known as bentonite), and the chemical industry. There are two main varieties of smectite, described in the following:

Sodium Smectite

This is the high-swelling form of smectite, which can adsorb up to 18 layers of water molecules between layers of clay. Sodium smectite is the preferred clay mineral for drilling muds, for creating a protective clay liner for hazardous waste landfills to guard against future groundwater contamination, and for preventing seepage of groundwater into residential basements. Sodium smectite will retain its water-tight properties so long as the slurry is protected from evaporation of water, which would cause extensive mud cracks. As a drilling mud, sodium smectite mixed with water to form a slurry which performs the following functions when drilling an oil or water well:

1. Lubricates the drill bit to prevent premature wear,
2. Prevents the walls of the drill hole from collapsing inwards,
3. Suspends the rock cuttings inside the dense mud so that the mud may pumped out of the drill hole, and
4. When the dense mineral barite is added to drilling mud, it prevents blow-outs caused by internal pressure encountered during deep drilling. Sodium smectite is also used as commercial clay absorbent to soak up spills of liquids.

Calcium smectite

The *low-swelling* form of smectite adsorbs less water than does sodium smectite, and costs less. Calcium smectite is used locally for drilling muds. Much of the domestic supplies of calcium smectite are mined from the state of Georgia.

Attapulgite: This mineral actually resembles the amphiboles more than it does clay minerals, but has a special property that smectite lacks - as a drilling fluid, it stable in salt water environments. When drilling for offshore oil, conventional drilling mud falls apart in the presence of salt water. Attapulgite is used as a drilling mud in these instances. Incidentally, attapulgite is the active ingredient in the current formula of Kaopectate.

GENESIS OF CLAY MINERALS

The silicate clays are developed from the weathering of a wide variety of minerals by the two distinct process

1. **Alteration** - A slight physical and chemical alteration of certain primary minerals. Changes in particle size. Alteration of muscovite mica to fine grained mica is the good example. As weathering occurs muscovite mineral is broken down in size to the colloidal range, part of the potassium is lost and some silicon is added from weathering solutions. Net result is the less rigid crystal structure and an electronegative charge. The fine mica colloid has a 2:1 structure, only have been altered in this process

2. **Recrystallization** - Decomposition of primary minerals with subsequent recrystallization of certain of their products in to the silicate clays

Complete breakdown of clay structures and re-crystallization of clay minerals from product of this breakdown. It is the result of much more intense weathering than that required for alteration. Formation of Kaolinite (1;1) from solutions containing soluble aluminum and silicon that came from the breakdown of the primary minerals having 2;1 type structure.

Relative stages of weathering

Weathering

The contact of rocks and water produces clays, either at or near the surface of the earth.

Rock + Water → Clay

The CO_2 gas can dissolve in water and form carbonic acid, which will become

hydrogen ions H^+ and bicarbonate ions, and make water slightly acidic.

$CO_2 + H_2O \rightarrow H_2CO_3 \rightarrow H^+ + HCO_3^-$

The acidic water will react with the rock surfaces and tend to dissolve the K ion and silica from the feldspar. Finally, the feldspar is transformed into kaolinite.

Feldspar + hydrogen ions + water - clay (kaolinite) + cations, dissolved silica

$$2KAlSi_3O_8 + 2H^+ + H_2O \; \square \; Al_2Si_2O_5(OH)_4 + 2K^+ + 4SiO_2$$

Fine grained micas and magnesium rich chlorites represent earlier weathering stages of the silicates and kaolinite and ultimately iron and aluminum oxides the most advanced stages.

CONCEPT OF SILICATE CLAYS

Size and Chemical Composition:

The chemical analysis of clay indicates the presence of silica, alumina, iron and combined water. These make up from 90-98 per cent of the colloidal clay. The soil colloidal matter contains plant nutrients like Ca, Mg and K etc. Clay is a mixture of hydrated aluminoferro silicates of varying composition mixed in some cases with an excess of sesquioxides or silica.

The term "clay" has three meanings in soil usage:

1. It is a particle fraction composed of any particles less than 2 microns ($< 2â\mu$) in effective diameter,
2. It is a name for minerals of specific composition; and
3. It is a soil textural class. Many materials of the clay size fraction, such as gypsum, carbonates or quartz are small enough to be classified as "clay" on the basis of size but are not "clay minerals". Sometimes some clay minerals have a size of 4 or 5 p (double the upper size limit of the "clay size fraction").

Shape:

Silicate clay minerals have been examined by electron microscope and found that the particles are laminated made up of layers of plates or flakes or even rods. Each clay particle is made up of a large number of plates like structural units.

The different units or flakes of clay minerals are held together with varying degrees of force depending upon the nature of the clay mineral. The edges of some clay particles are clean cut and others are frayed or fluffy. In all cases, clay minerals are developed more in the horizontal axis than that of vertical axis.

Surface Area

The surface area of a clay particle is usually defined as the area of the particle that is accessible to ions or molecules when the clay is in an aqueous solution. All clay particles (finer fraction of soil) must expose a large amount of external surface.

In some clay there are extensive internal surfaces as well. This internal exists between the plate like crystal units that make up each particle (Fig. 1). So the large surface area of clay colloids is not only due to its fineness but also its plate-like structure.

Surface areas of clay particles can be measured by using cetylpyridinium bromide (dissolving in water) for fully dispersed clay suspensions. Surface area for clays like sodium montmorillonite 700-800 m²/g; vermiculites and some mixed layer clays 300-500 m²/g; micaceous clays 100-300 m²/g; and kaolinitic clays 5-100 m²/g.

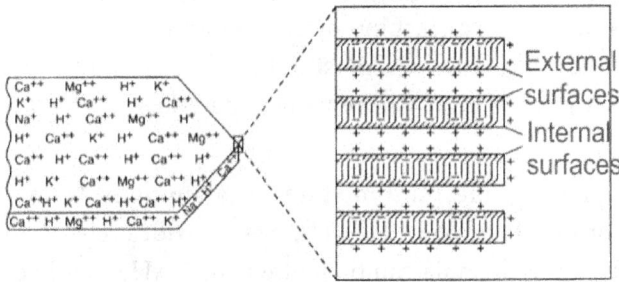

Fig 1. Internal and External surfaces of a plate like clay crystal unit

Amorphous clays (oxygen and other atoms less regularly oriented) have surface areas between 100 and 500 m²/g.

Electronegative Charge:

Clay micelles (micro cells) carry negative charges and so a number of oppositely charged ions (cations) are attracted to each colloidal clay crystal. The colloidal clay particles have inner ionic layer (surfaces of highly negative charge) and the outer ionic layer (cations swarming layer).

Adsorbed Cations

Clay micelles adsorb a number of cations-humid, arid and semiarid regions colloids-cations are H^+, Al^{3+}, Ca^{2+}, $Mg^{/+}$, Na^+ and K^+. The amount of these cations held by clay varies with its kind. Cations adsorbed (if dominant) on the clay colloids very oftenly determines the physical and chemical properties of the soil and thereby influence the plant growth.

Example

Alkali soil dominated by sodium on the surface of the clay micelle causes poor physical condition of the soil.

AMORPHOUS CLAYS

They are mixtures of silica and aluminium that have not formed well-oriented crystals. Technically they are not minerals because they lack crystallinity. These clays occur where large amounts of weathered products existed but have not had the conditions or time for good crystal formation. Amorphous clays are common in soils forming from volcanic ash.

Their properties are inconsistent, such as having high positive charges (high anion exchange capacities) or even high cation exchange capacity. Because almost all of their charge is from hydroxyl (OH^-) ions, which can gain a positive ion or lose the H^+ attached, these clays have a variable charge that depends on how much H^+ is in solution (the soil acidity).

SESQUIOXIDE CLAYS (METAL OXIDES AND HYDROUS OXIDES):

Under conditions of extensive leaching by rainfall and long-time intensive weathering of minerals in humid warm climates, most of the silica and much of the aluminium are dissolved and slowly leached away. The remanant materials, which have lower solubilities, are sesquioxides.

Sesquioxides (metal oxides) are mixtures of aluminium hydroxide, $Al(OH)_3$, and iron oxide, Fe_2O_3 or iron hydroxide, $Fe(OH)_3$. Sesquioxides refer to the clays of iron and aluminium because their formula can be written $Al_2O_3.xH_2O$ and $Fe_3O_3; xH_2O$, one and one-half times more oxygen than of Al or Fe. These clays may be amorphous or crystalline and do not swell. They are not sticky and do not behave as like that of the silicate clays.

STRUCTURE OF SILICATE CLAYS

Most of the silicate clays are made up of planes of oxygen atoms with silicon and aluminium atoms holding the oxygen together by ionic bonding, which is the attraction of positively and negatively charged atoms. Three or four planes of oxygen atoms with intervening silicon and aluminium ions make up a layer.

One clay particle is composed of many layers stacked like a deck of cards. Most silicate clays are aluminosilicates (aluminium and silicon components of the clay structure).

One silicon atom surrounded by four oxygen atoms forming silica tetrahedron (because of its four sided configuration) and the unit has tetrahedral coordination. The planes of oxygen held together by Si^{4+} are tetrahedrally oriented and are referred to as a silica tetrahedral sheet. The aluminium octahedron is an eight sided building block consisting of central aluminium atom surrounded by six hydroxyls or oxygen.

Large numbers of aluminiumoctahedra, bound to each other by shared oxygen atoms in an octahedral layer, are arranged in a plane forming aluminium octahedral sheet. One silica sheet per one aluminium sheet is a 1: 1 lattice A 2: 1 lattice has 2 silica sheets per 1 aluminium sheet.

Different combinations of these two general structural units (tetrahedral and octahedral sheets) form the structures of the various layer silicates like mica, vermiculite, montmorillonite, chlorite, kaolinite and other interstratified and intergradient layer silicates.

CLASSIFICATION OF SILICATE CLAYS

On the basis of the number and arrangement of silica and alumina sheets, silicate clays may be classified into four different groups:

    i. 1: 1 type minerals;

    ii. 2: 1 type minerals (expand between crystal units);

iii. 2: 1 type non-expanding minerals; and

iv. 2: 2 type minerals.

1: 1 Type Minerals:

The most important mineral in this type, commonly found in soils, is kaolinite whose structure is shown in the Fig. 2. The chemical composition of kaolinite is $Si_4Al_4O_{10}(OH)_8$. The two sheets of each crystal unit of kaolinite are held together by oxygen atoms which are mutually shared by the silicon and aluminium atoms in their respective sheets.

These units are held together very rigidly by hydrogen bonding (-H-) to the oxygen plane of the adjacent layer. So the lattice is fixed and kaolinite mineral does not allow water to penetrate between the layers and has almost no plasticity, cohesion, shrinkage and swelling properties. Besides kaolinite, there are other minerals in this type namely halloysite, anauxite and dickite.

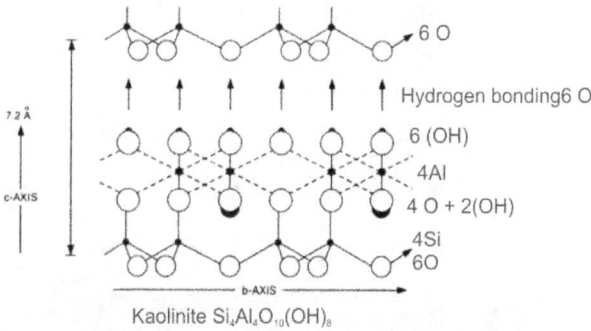

Fig. 2 Structure of Kaolinite (1 : 1 layer silicate) mineral

2: 1 Type Minerals (Expanding Lattice Type):

There are some important minerals in this type which includes montmorillonite (smectitic group), vermiculite and other smectitic group of minerals like beidellite, nontronite and saponite etc. Montmorillonite is best known example of this type of minerals and its structure is shown (Fig. 3).

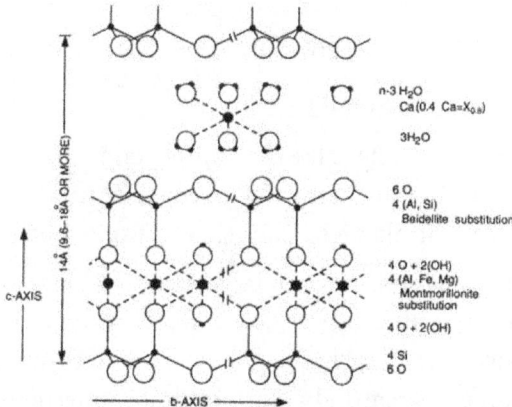

Fig. 3. Structure of montmorillonite (2 : 1 layer silicate) mineral

The flake like crystals of this mineral are composed of 2: 1 type crystal units (Fig. 3). These crystal units are loosely held together by very weak oxygen to oxygen linkages. Water molecules as well as cations are attracted between crystal units, causing expansion of the crystal lattice.

The spacing (C-Axis) of the layers ranges from 1.2-1.8 nanometers and is variable with the exchangeable cation species and the degree of inter layer solvation.

Montmorillonites are the swelling and sticky clays. Internal surface, cohesion and plasticity of this mineral are also very high. The size of this mineral is very small (0.01-1. Op m or microns). On drying it shows shrinkage and as a result wide cracks usually form as soils dominated this type of mineral. The dry aggregates or clods are very hard, making such soils difficult to till.

Vermiculite clays are common in most soils. The structure of vermiculite (Fig. 4) is similar to that of hydrous mica structure but also has the layers held more weakly together by hydrated magnesium (six water molecules in octahedral coordination with Mg^{2+}) rather than tightly held by potassium ions (K^+). Thus, vermiculite has swelling but not as much as montmorillonite. It has a high cation exchange capacity.

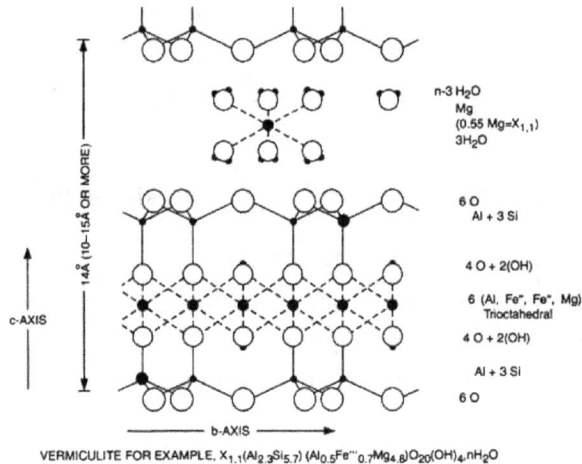

VERMICULITE FOR EXAMPLE, $X_{1.1}(Al_{2.3}Si_{5.7}) (Al_{0.5}Fe'''_{0.7}Mg_{4.8})O_{20}(OH)_4.nH_2O$

Fig. 4. Structure of Vermiculite clays

2: 1 Non-Expanding Type Minerals

In this group, hydrous mica or illite is the most important in soils. Like montmorillonite, illite has a 2: 1 type lattice. The structure of illite is presented diagrammatically in Fig. 5. However, about 15 per cent of silicon in silica sheets is substituted by aluminium.

The excess of negative charge is satisfied largely by potassium in the inter-lattice layers, thus making the lattice structure of the non-expanding type. Hence illite is relatively non-expansive. Thus hydrous mica has slight to moderate swelling. The properties of water adsorption, cation exchange and other physical properties lie in between those of kaolinite and montmorillonite type of minerals.

2: 2 Type Minerals

Chlorite (2: 2 or 2: 1: 1 layer silicates) occurs extensively in soils. Chlorites are basically silicates of magnesium with some iron and aluminium present and it is composed of alternate talc (similar to a montmorillonite crystal unit) and brucite $[Mg(OH)_2]$ layers (Fig. 6).

Chlorite mineral is similar to the unit lattice of vermiculite, except the hydrated Mg in vermiculite is a firmly bonded magnesium hydroxide octahedral sheet.

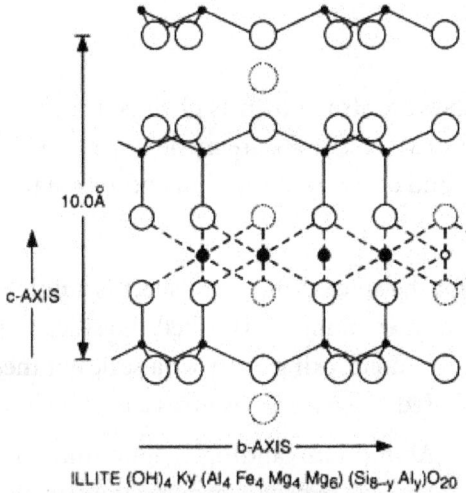

ILLITE $(OH)_4 Ky (Al_4 Fe_4 Mg_4 Mg_6) (Si_{8-y} Al_y)O_{20}$

Fig. 5 Structure of illite mineral

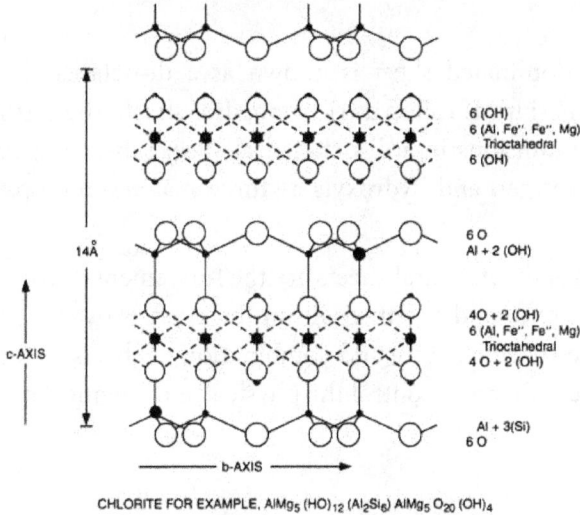

CHLORITE FOR EXAMPLE, $AlMg_5 (HO)_{12} (Al_2Si_6) AlMg_5 O_{20} (OH)_4$

Fig. 6 Structure of Chlorite mineral

Thus, a layer of chlorite has 2 silica tetrahedra, an aluminiumoctahedra and a magnesium octahedra sheet (2: 2 or 2: 1: 1). Chlorite does not swell on wetting and has low cation exchange capacities. It is almost non-expanding type of mineral because of its very little water adsorption.

GENESIS OF INDIVIDUAL CLAYS

General conditions for the formation of the various layer silicate clays and oxides of iron and aluminum. Fine-grained micas, chlorite, and vermiculite are formed through rather mild weathering of primary aluminosilicate minerals, whereas kaolinite and oxides of iron and aluminum are products of much more intense weathering. Conditions of intermediate weathering intensity encourage the formation of smectite. In each case silicate clay genesis is accompanied by the removal in solution of such elements as K, Na, Ca, and Mg.

Layer silicate clays

These important silicate clays are also known as phyllosilicates (Phyllon - leaf) because of their leaf-like or plate like structure. These are made up of two kinds of horizontal sheets. One dominated by silicon and other by aluminum and/or magnesium.

Silica tetrahedron

The basic building block for the silica-dominated sheet is a unit composed of one silicon atom surrounded by four oxygen atoms. It is called the silica tetrahedron because of its four-sided configuration. An interlocking array or a series of these silica tetrahedra tied together horizontally by shared oxygen anions gives a tetrahedral sheet.

Alumina octahedron: Aluminium and/or magnesium ions are the key cations surrounded by six oxygen atoms or hydroxyl group giving an eight sided building block termed octahedron. Numerous octahedra linked together horizontally comprise the octahedral sheet.

An aluminum-dominated sheet is known as a di-octahedral sheet, whereas one dominated by magnesium is called a tri-octahedral sheet. The distinction is due to the fact that two aluminum ions in a di-octahedral sheet satisfy the same negative charge from surrounding oxygen and hydroxyls as three magnesium ions in a tri-octahedral sheet.

The tetrahedral and octahedral sheets are the fundamental structural units of silicate clays. These sheets are bound together within the crystals by shared oxygen atoms into different layers. The specific nature and combination of sheets in these layers vary from one type of clay to another and control the physical and chemical properties of each clay.

14

SOIL AND THE CARBON CYCLE

Carbon (C) is one of the most common elements in the universe and found virtually everywhere on earth: in the air, the oceans, soil, and rock. Carbon is part of geologic history in rock and especially the ancient deposits that formed coal, oil and other energy sources we use today. Carbon is also an essential building block of life and a component of all plants and animals on the planet. It has unique bonding properties that allow it to combine with many other elements. These properties enable the formation of molecules that are useful and necessary to support life. The role of carbon in living systems is so significant that a whole branch of study is devoted to it: organic chemistry. Carbon that is not tied up in rock or deep in the oceans is constantly changing and moving. This process is called the carbon cycle. Soil holds the largest portion of active carbon on earth. Plants take carbon from the air and convert it to plant tissue, some of which returns to the soil as plant residue.

AGRICULTURE'S ROLE IN THE CARBON CYCLE

Carbon is critical to soil function and productivity, and a main component of and contributor to healthy soil conditions. Soil management plays a critical role in whether the carbon remains in the soil or is released to the atmosphere. Agricultural practices can impact both the amount and the composition of soil organic carbon and hence also the soil's physical, biological, and chemical condition, the combination of things that defines soil health. Farm practices that affect carbon therefore impact agricultural productivity and resilience (the soil's ability to deal with weather extremes) and the carbon cycle itself.

IMPORTANCE OF SOIL ORGANIC CARBON

While the agricultural sector has the ability to impact the carbon cycle on a large scale, often through the release of carbon, farmers have a vested interest in retaining and increasing soil organic carbon for individual fields because soil and yield tend to improve when the soil organic carbon level increases. Higher soil organic carbon promotes soil structure or tilth meaning there is greater physical stability. This improves soil aeration (oxygen in the soil) and water drainage and retention, and reduces the risk of erosion and nutrient leaching. Soil organic carbon is also important to chemical composition and biological productivity, including fertility and nutrient holding capacity of a field. As carbon stores in the soil increase, carbon is "sequestered", and the risk of loss of other nutrients through erosion and leaching is reduced. An increase in soil organic carbon typically results in a more stable carbon cycle and enhanced overall agricultural productivity, while physical disturbances of the soil can lead to a net loss of carbon into the surrounding environment due to formation of carbon dioxide (CO_2).

In soils, you can find carbon in both organic carbon compounds and inorganic carbon compounds. In most soils, carbon exists predominately in the form of soil organic carbon (SOC).

Soil organic carbon (SOC) is the main constituent of **soil organic matter (SOM)**. SOM is formed by the biological, chemical and physical decay of organic materials on the soil surface and below the ground. Basically, soil organic matter (SOM) is composed of anything that once lived, including:

- ◉ **Organic bits and pieces of plant and animal remains** in various stages of decomposition, sloughed off cells and tissues of soil organisms, and substances from plant roots and soil microbes.

- ◉ **Living soil microbes** (bacteria, fungi, archaea, nematodes and protozoa) and plant roots. If we weighed all of the organisms found in soil, soil microbes would comprise about 90-95% of that weight.

- ◉ **Humus,** a chemically stable type of organic matter composed of large, complex organic carbon compounds, minerals, and soil particles. Humus is resistant to further decomposition unless disturbed by a change in environmental conditions. If undisturbed, humus can store soil carbon for hundreds to thousands of years. This makes humus a very important carbon sink.

- ◉ **Charcoal** (biochar), incompletely burned plant material. Charcoal can remain undecomposed in the soil for decades to centuries.

The carbon balance within soil is controlled by carbon inputs from photosynthesis, carbon losses by respiration and carbon storage in humus

Examine the image pictured on the right showing the flow of carbon into and out of soil. If plants transfer more carbon into soil than is released via soil respiration, more carbon will get stored in humus via the process of **humification**.

For years, soil scientists and farmers have known that carbon can persist for long periods of time in carbon-rich humus. Humus is formed when soil organic material is degraded by soil microbes that reside in the soil. However, humus is a very complex substance and not fully understood. Current research indicates that the length of time soil carbon persists in humus and other SOM components depends on many ecosystem interactions between SOM and microbes, minerals, moisture and temperature. Scientists do not know yet how long this carbon will stayed stored in soil or if environmental disturbances will move large amounts into the atmosphere amplifying climate change.

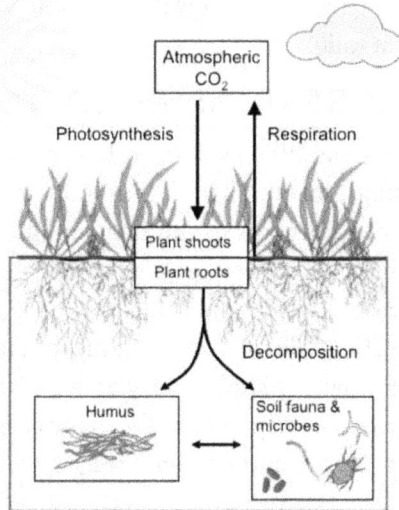

Fig. 1 Soil Carbon Storage: Carbon balance within the soil (brown box) is controlled by carbon inputs from photosynthesis and carbon losses by respiration.

Plant root exudation transfers a variety of complex organic carbon compounds such as acids, sugars and protein enzymes from the roots of trees and shrubs into the surrounding soil and into a symbiotic ecosystem of **mycorrhizal fungi.** These fungi get sugars from the plant in return for greatly enhancing the plant's ability to take up water and nutrients from the soil. Mycorrizal fungi eat the sugars and then deposit carbon-containing residue in the surrounding soil. Recent research indicates some types of these fungi (ectomycorrhizal fungi) can lead to up to 70% more carbon stored in soil.

Soil microbes move and transform carbon compounds and make nitrogen bioavailable to plants.

Some of the smallest organisms in both soil and the oceans have key roles in moving and transforming carbon compounds in their ecosystems. In soils, microbes directly and indirectly mediate about 90% of soil functions, such as:

- Decomposing dead matter into som;
- Respiring co_2 and methane (ch_4) to the surrounding soil and air;
- Making essential biogeochemical nutrients such as nitrogen compounds bioavailable to plants and other soil organisms;
- Storing carbon in soil humus; and
- Decomposing humus which releases CO_2 to the air via soil respiration.

All organisms in the Biosphere need nitrogen to build their DNA, RNA and protein molecules. Because plants transfer carbon into the soil via photosynthesis, the nitrogen cycle becomes critical to building strong healthy soil. Take a few minutes to examine the image of the nitrogen cycle above and then watch the Soil Microbes video. Then answer the discussion questions that follow.

ACTIVE SOIL MICROBES RESPIRE CO_2 AT GREATER RATES

Soil microbes have a busy lifestyle! When soil microbes are carrying out their life activities such as reproducing, eating, metabolizing, biosynthesizing, transforming nitrogen compounds, decomposing dead organisms, humus and SOM, they need lots of energy. Burning sugars via respiration provides that energy and releases CO_2 as a by-product. This CO_2 can be used by plants covering the soil surface for photosynthesis and/or drift up to the air to join other CO_2 molecules in the atmosphere. Several abiotic and biotic environmental variables can make microbes very active or can slow down that activity. How active or inactive microbes are depends on the many ecosystem interactions between microbes, minerals, SOM, moisture and temperature.

SOIL MICROBES, SOIL RESPIRATION AND CLIMATE CHANGE

When scientists discuss global climate change, they often focus on the amount of carbon in the atmosphere and vegetation. In reality, soil contains more carbon than Earth's vegetation and atmosphere combined. For this reason, even a minor change in soil carbon could have major implications for Earth's atmosphere and climate.

Scientists agree that they have many unanswered questions about how soil will respond to climate change and more research needs to be done. Key research questions scientists are exploring are:

- How will soil and soil microbes respond to a warming climate?

- Will a warming climate result in higher amounts of carbon transferring from soils to the atmosphere as CO_2 and methane (CH_4), both greenhouse gases that could create additional warming.

To find answers to these important questions, scientists need to know more about how soil microbes respond to changes in climate environmental variables such as temperature and moisture. To that end, you will design and carry out an experiment to test the effect of temperature on soil respiration rates.

LABORATORY INVESTIGATION INSTRUCTIONS:

1. Prepare your bottle. Use scissors or safety blade to start a cut about 4-5 cm from the top of the bottle. Then, cut approximately 8/10 of the way around the bottle. DO NOT cut all the way around. You want to be able to fold the top of the bottle back as you fill the bottle with soil and water.

2. Test the soil for N-P-K nutrients (optional)
3. Put the soil and sugar water into the bottle in increments:

- Fill the bottles approximately 1/3 to the top, add 20 ml of water and then gently tap the bottle on the counter several times to compact the soil.

- Repeat the above. Add more soil to 2/3 full and add 20 ml of sugar water and tap to compact.

- Fill bottle with soil up to the cut rim. Add 20 ml of sugar water and tap.

- Close and seal the bottle cut with duct tape. Add more soil through the mouth of the bottle to bring the level of soil up to 1 cm below the mouth of the bottle where the rubber stopper will go. Add 10 more ml of sugar water.

- Insert the #4 one-hole stopper gas delivery apparatus into the bottle top. Insert the tubing into the collecting vessel that contains 50 ml of limewater. Make sure the end of the tubing is in the limewater.

- Use the aluminum foil to cover the collecting vessel and wrap tightly to make sure no water from the lime water can evaporate and that no CO_2 from the surrounding air can get into the limewater.

- Put your bottle, soil and limewater set-up under the different temperature conditions your class has identified.

- Use a Celsius thermometer to take the ambient temperature for each different temperature condition.

- Observe over the next 2-3 days. Look for tiny "flakes" of white, chalky calcium carbonate and/or cloudiness forming in the limewater.

- Record your results in your lab notebook if your teacher requires you to keep one.

SOIL ORGANIC MATTER CYCLING

Soil type, climate and management influence organic matter inputs to soil and its turnover or decomposition. Rainfall is a major driver of plant growth (biomass) and biological activity which results in the decomposition of organic matter that enters soil. The different fractions of SOM (dissolved, particulate, humus and resistant) turn over at vastly different rates (Fig. 2). Furthermore, SOM cycles continuously between living, decomposing and stable fractions in the soil (Fig. 3).

1. Inputs: plants and animals become part of the SOM as they die or create by-products.
2. Transformation: soil organisms break-up and consume organic matter, creating different forms of organic residues. For example, fresh plant residues are broken into smaller pieces (<2mm) and become part of the particulate organic matter fraction. This material decomposes further and a smaller amount of more biologically stable material enters the humus pool.

3. Nutrient release: nutrients and other compounds not required by microorganisms are released and are then available to plants.
4. Stabilising organic matter: as the organic residues decompose, they become more resistant to further change.

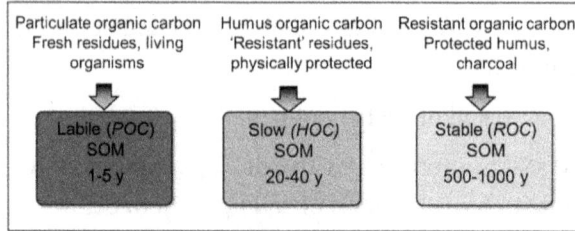

Particulate organic carbon Fresh residues, living organisms	Humus organic carbon 'Resistant' residues, physically protected	Resistant organic carbon Protected humus, charcoal
Labile (POC) SOM 1-5 y	Slow (HOC) SOM 20-40 y	Stable (ROC) SOM 500-1000 y

Fig. 2 Different fractions of soil organic matter decompose in the soil over different time frames

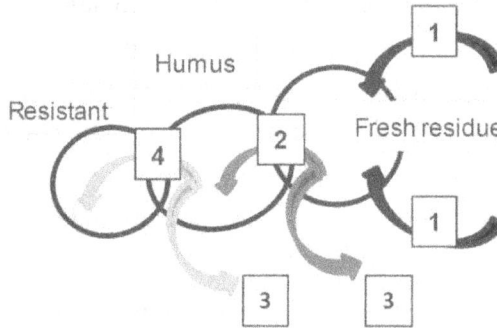

Fig. 3 Organic matter transforms from one form to another as it decomposes and cycles into different soil organic fractions

MANAGEMENT PRACTICES FOR C SEQUESTRATION

With agricultural productivity so dependent on soil organic carbon and carbon cycling, how can we best manage fields to enhance soil organic carbon levels while also reducing carbon loss into the atmosphere?

The ability of agricultural fields to sequester carbon (capture and storage of carbon that would otherwise be lost to the environment) depends on several factors including climate, soil type, type of crop or vegetation cover, and management practices. Employing farming practices that reduce disturbance of the soil (less aeration from tillage helps protect carbon), combined with practices that bring additional carbon to the soil, will allow for carbon sequestration over time. Such practices include implementation of conservation tillage (no-till, zone-till, minimum-till, shallow mixing or injection for manure applications), retaining crop residues, including cover crops in crop rotations, adding organic nutrient sources such as manure and compost, and including perennial crops in crop rotations (Table 1). Their implementation may slow or even reverse the loss of carbon from agricultural fields, improve nutrient cycling and reduce nutrient loss.

Table 1. Management practices that can increase soil organic carbon and reduce carbon loss into the atmosphere.

Management practices	Functions and explanation Conservation tillage practices Conservation tillage practices including no-till management aid in storing soil organic carbon, keeping the physical stability of the soil intact. When reduced-till systems are combined with residue management and manure management, soil organic carbon can increase over time.
Crop residue management	Returning crop residue to the soil adds carbon and helps to maintain soil organic matter.
Cover crops	Cover crops can increase soil carbon pools by adding both root and above ground biomass. Covers also reduce the risk of soil erosion and the resulting loss of carbon with soil particles. Cover crops also enhance nutrient cycling and increase soil health over time.
Manure and compost	Adding organic amendments such as manure or compost can directly increase soil carbon, and also result in increased soil aggregate stability. This enhances the biological buffering capacity of the soil, resulting in greater yields and yield stability over time.
Crop selection	Perennial crops eliminate the need for yearly planting and increase soil organic carbon by root and litter decomposition post-harvest. Crops with greater root mass in general add to root decomposition and physically bond aggregates together. Using high residue annual crops can also help reduce net carbon loss from cropping systems.

15

SOIL COLLOIDS

INTRODUCTION

The extremely small, colloidal particles (smaller than 0.001 mm) of clay and humus control many important chemical and physical properties of the soil. This portion of the soil is often called the "active fraction". The small size of colloids results in a large surface area per unit weight, and their ionic structure results in a net electrical charge. The type, amount, and mineralogy of colloids will strongly influence most land management decisions. For example a soil that is 40% of clay that primarily consists of smectite (a 2:1 shrink-swell clay) could have limitations for constructing roads, or building foundations due to the shifting of the soil as the soil wets and dries.

Such a soil could be highly productive for row crop agriculture though, due to the high amount of charge that facilitates the retention of nutrients like Ca^{2+}, K^+, Mg^{2+}, etc. On the contrary, a soil, such as an Oxisol that has 80% clay has colloids that are primarily aluminum and iron oxides, which do not shrink or swell, and have a low amount of charge. Thus, the soil would be well suited for building foundations. However, the high phosphorus fixation capacity, and the limited ability to retain base cations limit the productivity of the soil for row crop agriculture.

Ion exchange is one of the most significant features of the clay and humus fractions. The capacity of the particles to attract or adsorb cations is called the cation exchange capacity. This ability allows the soil to serve as a storehouse of plant nutrients like potassium, calcium, and magnesium. This reactive exchange capacity also permits the soil to serve as a filter or treatment medium for land application of waste materials.

DETERMINING CATION EXCHANGE CAPACITY

The cation exchange capacity (quantity of cations a soil can adsorb per unit weight, CEC) can be determined using a simple displacement process (Fig. 1). In step 1, a soil sample is first saturated with a simple cation like NH_4^+ so all the negative charge sites are occupied by NH_4^+. In step 2, excess NH_4^+ (i.e., not on exchange sites) is removed by leaching with ethyl alcohol. In step 3, another cation such as Ba^{2+} is used to displace all the NH_4^+. The NH_4^+ is collected in the filtrate and measured. The quantity of NH_4^+ collected from the sample is the quantity of cations that the soil can hold, i.e. CEC.

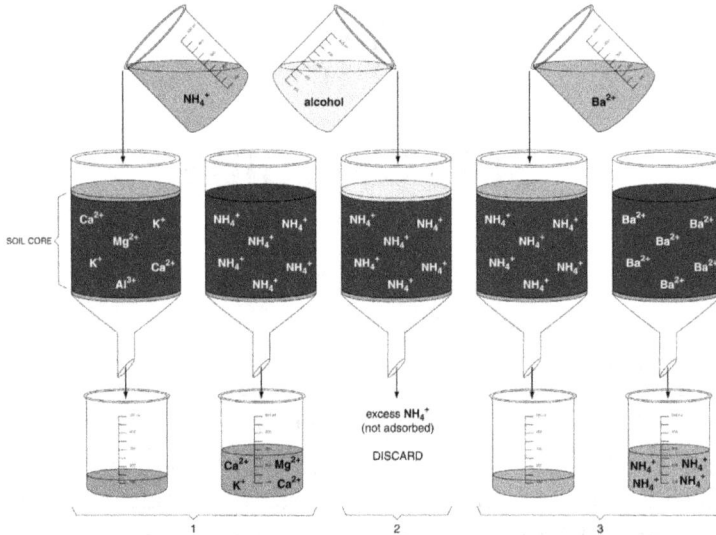

Fig. 1. Series of chemical extractions to determine CEC

Many cations could be used in step 1 of the displacement method. Soil laboratories often use the ammonium ion. However, in this activity, H_+ will be used as the saturating cation. Therefore, we will determine the amount of extracted H^+ in the filtrate. Then, using that amount, you will calculate the CEC for each soil sample.

Calculating CEC uses the concept of moles of charge. In chemistry, one mole of an element is the quantity of the element with a weight in grams numerically equal to its atomic weight. For example, the atomic weight of K is 39.1, so one mole of K weighs 39.1 g. One mole of any element contains 6.02×10^{23} atoms of the element. Similarly, one mole of charge is 6.02×10^{23} charges. If K exists as a cation in solution (K^+), then a solution containing a mole of K would contain 6.02×10^{23} positive charges.

The number of positive charges in the filtrate is thus equal to the number of negative charges on the exchange sites in the soil sample. Therefore, your task is to determine the number of charges in the filtrate. Each H^+ ion has one positive charge, so by determining the amount of H^+ in the filtrate, you can determine the number of positive charges in the filtrate and by extension, the number of negative charges on the exchange sites in the soil sample.

TYPES OF SOIL COLLOIDS

There are four major types of colloids present in soil

1. Layer silicate clays
2. Iron and aluminum oxide clays (sesquioxide clays)

3. Allophane and associated amorphous clays
4. Humus.

Layer silicate clays, iron and aluminum oxide clays, allophane and associated amorphous clays are inorganic colloids while humus is an organic colloid.

Layer silicate clays

These are most important silicate clays and are known as phyllosilicates (Phyllon – leaf) because of their leaf-like or plate like structure. They are comprised of two kinds of horizontal sheets. One dominated by silicon and other by aluminum and/or magnesium.

Silica Tetrahedron

The basic building block for the silica-dominated sheet is a unit composed of one silicon atom surrounded by four oxygen atoms. It is called the silica tetrahedron because of its four-sided configuration. An interlocking array or a series of these silica tetrahedral tied together horizontally by shared oxygen anions gives a tetrahedral sheet.

Alumina Octahedron

Aluminium and/or magnesium ions are the key cations in the second type of sheet. An aluminium (or magnesium) ion surrounded by six oxygen atoms or hydroxyl group gives an eight sided building block termed octahedron. Numerous octahedra linked together horizontally comprise the octahedral sheet. An aluminum dominated sheet is known as a dioctahedral sheet, whereas one dominated by magnesium is called a trioctahedral sheet. The distinction is due to the fact that two aluminum ions in a dioctahedral sheet satisfy the same negative charge from surrounding oxygen and hydroxyls as three magnesium ions in a trioctahedral sheet.

The tetrahedral and octahedral sheets are the fundamental structural units of silicate clays. They, in turn, are bound together within the crystals by shared oxygen atoms into different layers. The specific nature and combination of sheets in these layers vary from one type of clay to another and largely control the physical and chemical properties of each clay.

Types of Silicate Clay Minerals: On the basis of the number and arrangement of tetrahedral (silica) and octahedral (alumina-magnesia) sheets contained in the crystal units or layers, silicate clays are classified into three different groups

A. 1:1 Type Minerals

The layers of the 1:1-type minerals are made up of one tetrahedral (silica) sheet combined with one octahedral (alumina) sheet-hence the terminology. In soils, kaolinite is the most prominent member of this group, which includes hallosite, nacrite, and dickite.

The tetrahedral and octahedral sheets in a layer of a kaolinite crystal are held together tightly by oxygen atoms, which are mutually shared by the silicon and aluminum cations in their respective sheets. These layers, in turn, are held together by hydrogen bonding. Consequently, the structure is fixed and no expansion ordinarily occurs between layers when the clay is wetted.

Cations and water do not enter between the structural layers of a 1:1 type mineral particle. The effective surface of kaolinite is thus restricted to its outer faces or to its external surface area. Also, there is little isomorphous substitution in this 1:1 type mineral. Along with the relatively low surface area of kaolinite, this accounts for its low capacity to adsorb cations.

Kaolinite crystals usually are hexagonal in shape. In comparison with other clay particles, they are large in size, ranging from 0.10 to 5 um across with the majority falling within the 0.2 to 2 um range. Because of the strong binding forces between their structural layers, kaolinite particles are not readily broken down into extremely thin plates.

Kaolinite exhibits very little plasticity (capability of being molded), cohesion, shrinkage, and swelling.

B. 2:1-Type Minerals

The crystal units (layers) of these minerals are characterized by an octahedral sheet sandwiched between two tetrahedral sheets. Three general groups have this basic crystal structure.

      i. Expanding type: smectites and vermiculite
     ii. Non-expanding type: mica (illite)

Expanding Minerals: The smectite group is noted for interlayer expansion, which occurs by swelling when the minerals are wetted, the water entering the interlayer space and forcing the layers apart. Montmorillonite is the most prominent member of this group in soils, although beidellite, nontronite, and saponite are also found.

The flake-like crystals of smectites (e.g., Montmorillonite) are composed of an expanding lattice 2:1-type clay mineral. Each layer is made up of an octahedral sheet sandwiched between two tetrahedral (silica) sheets. There is little attraction between oxygen atoms in the bottom tetrahedral sheet of one unit and those in the top tetrahedral sheet of another. This permits a ready and variable space between layers, which is occupied by water and exchangeable cations. This internal surface far exceeds the surface around the outside of the crystal. In montmorillonite magnesium has replaced aluminum in some sites of the octahedral sheet. Likewise, some silicon atoms in the tetrahedral sheet may be replaced by aluminum. These substitutions give rise to a negative charge.

These minerals show high cation exchange capacity, marked swelling and shrinkage properties. Wide cracks commonly form as smectite dominated soils (e.g., Vertisols) are dried. The dry aggregates or clods are very hard, making such soils difficult to till.

Vermiculites are also 2: 1 type minerals in that an octahedral sheet occurs between two tetrahedral sheets. In most soils vermiculites, the octahedral sheet is aluminum dominated (dioctahedral), although magnesium dominated (trioctahedral) vermiculites are also common. In the tetrahedral sheet of most vermiculite, considerable substitution of aluminum for silicon has taken place. This accounts for most of the very high net negative charge associated with these minerals.

Water molecules, along with magnesium and other ions, are strongly adsorbed in the interlayer space of vermiculites. They act primarily as bridges holding the units together rather than as wedges driving them apart. The degree of swelling is, therefore considerable less for vermiculites than for smectites. For this reason, vermiculites are considered limited-expansion clay minerals, expanding more than kaolinite but much less than the smectites.

The cation exchange capacity of vermiculites usually exceeds that of all other silicate clays, including montmorillonite and other smectites, because of very high negative charge in the tetrahedral sheet. Vermiculite crystals are larger than those of the smectites but much smaller than those of kaolinite.

Non-expanding minerals: Micas are the type minerals in this group. (e.g.) Muscovite and biotite, weathered minerals similar in structure to these micas are found in the clay fraction of soils. They are called fine-grained micas. Like sanctities, fine-grained micas have a 2:1-type crystal. However, the particles are much larger than those of the smectites. Also, the major source of charge is in-the tetrahedral sheet where aluminum atoms occupy about 20% of the silicon sites. These results in a high net negative charge in the tetrahedral sheet, even higher than that found in vermiculites, to satisfy this charge, potassium ions are strongly attracted in the interlayer space and are just the right size to fit into certain spaces in the adjoining tetrahedral sheets. The potassium thereby acts as a binding agent, preventing expansion of the crystal. Hence, fine-grained micas are quite non expansive.

The properties such as hydration, cation adsorption, swelling, shrinkage, and plasticity are much less intense in fine-grained micas than in smectites. The fine grained micas exceed kaolinite with respect to these characteristics, but this may be due in part to the presence of interstratified layers of smectite or vermiculite. In size, too, fine-grained mica crystals are intermediate between the smectities and kaolihites.

Their specific surface area varies from 70 to 100 m^2/g, about one eighth that for the smectites.

2:1:1 Type Minerals

This silicate group is represented by chlorites, which are common in a variety of soils. Chlorites are basically iron magnesium silicates with some aluminum present. In a typical chlorite clay crystal, 2:1 layers, such as in vermiculites, alternate with a magnesium-dominated trioctahedral sheet, giving a 2:1:1 ratio. Magnesium also dominates the trioctahedral sheet in the 2:1 layer of chlorites. Thus, the crystal unit contains two silica tetrahedral sheets and two magnesium-dominated trioctahedral sheets giving rise to the term 2:1:1 or 2:2-type structure.

The negative charge of chlorites is about the same as that of fine-grained micas considerably less than that of the smectites or vermiculites. Like fine micas, chlorites may be interstratified with vermiculites or smectites in a single crystal. Particle size and surface area for chlorites are also about the same as for fine grained micas. There is no water adsorption between the chlorite crystal units, which accounts for the non expanding nature of this mineral.

Mixed and interstratified layers

Specific groups of clay minerals do not occur independently of one another. In a given soil, it is common to find several clay minerals in an intimate mixture. Furthermore, some mineral colloids have properties and composition intermediate between those of any two of the well defined minerals described. Such minerals are termed mixed layer or interstratified because the individual layers within a given crystal may be of more than one type. Terms such as "chlorite-vermiculite" and "fine-grained mica- smectite" are used to describe mixed-layer minerals. In some soils, they are more common than single-structured minerals such as montmorillonite.

Iron and aluminum oxide clays (sesquioxide clays)

Under conditions of extensive leaching by rainfall and long time intensive weathering of minerals in humid warm climates, most of the silica and much of the alumina in primary minerals are dissolved and slowly leached away. The remnant materials, which have lower solubility, are sesquioxides. Sesquioxides (metal oxides) are mixtures of aluminum hydroxide, $Al(OH)^3$, and iron oxide, Fe_2O_3, or iron hydroxide, $Fe(OH)^3$.

The Latin word sesqui means one and one- half times, meaning one and one-half times more oxygen than Al and Fe. These clays can grade from amorphous to crystalline.

Examples of iron and aluminum oxides common in soils are gibbsite ($Al_2O_3.3H_2O$) and geothite ($Fe_2O_3.H_2O$).

Less is known about these clays than about the layer silicates. These clays do not swell, not sticky and have high phosphorus adsorption capacity

Allophane and other Amorphous Minerals

These silicate clays are mixtures of silica and alumina. They are amorphous in nature. Even mixture of other weathered oxides (iron oxide) may be a part of the mixture. Typically, these clays occur where large amount of weathered products existed. These clays are common in soils forming from volcanic ash (e.g., Allophane). These clays have high anion exchange capacity or even high cation exchange capacity. Almost all of their charge is from accessible hydroxyl ions (OH-), which can attract a positive ion or lose the H+ attached. These clays have a variable charge that depends on H+ in solution (the soil acidity).

Humus (Organic Colloid)

Humus is amorphous, dark brown to black, nearly insoluble in water, but mostly soluble in dilute alkali (NaOH or KOH) solutions. It is a temporary intermediate product left after considerable decomposition of plant and animal remains. They are temporary intermediate because the organic substances remain continue to decompose slowly.

The humus is often referred to as an organic colloid and consists of various chains and loops of linked carbon atoms. The humus colloids are not crystalline. They are composed basically of carbon, hydrogen, and oxygen rather than of silicon, aluminum, iron, oxygen, and hydroxyl groups.

The organic colloidal particles vary in size, but they may be at least as small as the silicate clay particles. The negative charges of humus are associated with partially dissociated enolic (-OH), carboxyl (-COOH), and phenolic groups; these groups in turn are associated with central units of varying size and complexity.

PROPERTIES OF SOIL COLLOIDS

1. Colloidal particles are always in motion because of charge particles.
2. Colloidal particles are transformed from a liquid into a soft semisolid or solid mass by adding an opposite charged ion.
3. Colloidal particles have ability to absorbed gases, liquid and solid from their suspension.
4. Colloidal particles never pass through a semipermeable membrane.
5. Colloidal particles have the properties of cohesion and adhesion.

IMPORTANCE OF SOIL COLLOIDS

Soil colloids are important because their surfaces attract soil nutrients dissolved in soil, water as positively charged mineral ions, or cations. Some cations are needed for plant growth, including calcium (Ca++), Magnesium (Mg ++), Potassium (K+), and sodium (Na+). They need to be dissolved in a soil-water solution to be available to plants when they are in close contact with root membranes.

The fertility of the soil-water solution for plants is based on the capability of the soil to hold and exchange cations; this is referred to as the cation-exchange capacity. Without soil colloids, most vital nutrients would be leached out the soil by percolating water and carried away in streams.

The relative amount of bases held by a soil determines the base status of the soil. A high base status means that the soil holds an abundant supply of base cations necessary for plant growth. If soil colloids hold a small supply of bases, the soil is of low base status and is, therefore, less fertile. Humus colloids have high soil fertility.

Acid ions have the ability to replace the nutrient bases sticking to the surfaces of the soil colloids. As the acid ions force out the bases and build up, the bases are released into the soil solution. The bases are then gradually washed downward below, rooting level, weakening soil fertility. When this happens, the soil acidity is increased. Aluminium ions (Al+++), which are not plant nutrients, also have the capability to display base cations, reducing base status and soil fertility.

NATURE OF COLLOIDS:

Soil colloidal are two kinds:

1. Inorganic (minerals) and
2. Organic (humus).

The two together form the colloidal complex of the soil. In almost all soils the inorganic colloidal form a major portion of the colloidal complex. On the other hand, in peat soils, it consists almost entirely of organic colloids. Colloidal particles float in a medium and do not tend to settle. Some large colloidal particles may settle very slowly. Colloids are referred as the dispersed systems.

The substance in solution is termed as the dispersed phase while the medium in which the particles are dispersed is called the dispersion medium. The commonest colloids are those containing minute solid or liquid particles suspended in liquid or gas medium. Soils formed in tropical and semi-tropical regions the whole of the colloidal complex consists almost entirely of inorganic colloids.

Whereas soil formed in temperate regions usually contain more organic colloids than those formed in tropical and sub-tropical regions. In a broad way, two groups of clay are recognisedâ€" silicate clay as characteristic of temperate regions, and the iron and aluminium hydrous oxide clays found in tropical and semitropical.

CHEMICAL COMPOSITION AND STRUCTURE OF COLLOIDS:

The constitution of colloids are inorganic and organic:

Inorganic Colloids

The chemical analysis of clay indicates the presence of four main constituents; silica, alumina, iron and combined water. These make up from 90 to 98 per cent of the colloidal clay. The colloidal matter of soil contains a higher proportion of important plant nutrients such as Mg^{++}, Ca^{++} and K^+.

The shape of the individual particles is plate or flake-like (Fig. 2). Clay colloids are negatively charged (anions) and therefore attract a large number of positively charged ions (cations). The minute clay colloids particles, referred as micelles (micro cells), ordinarily carry negative charges.

Outer layer of positive ions   Inner layer of negetive ions

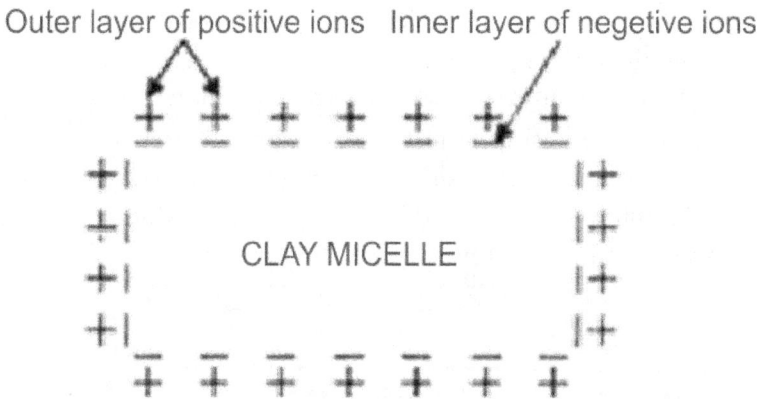

Fig. 2. Diagrammatic represntaion of colloidal clay crystals

Structure (constitution) of colloidal clay minerals are of two type:

a. The two-layer typesl (1:1 type) consists of one layer of silicon and oxygen atoms $(SiO)_2$ and other layer of aluminium and oxygen atoms (AI_2O_3), all in definite arrangement. Example: Kaloinite clay (Fig. 3). In this type of structure, there is non- expanding space between the sheets for the activity, thus, cation exchange capacity is low in kaolinite clay

b. Three-layer (2 : 1 type) clay crystal have two outside layers made of silicon and oxygen (SiO_2) and the middle layer of aluminium and oxygen (Al_2O_3). Montmorillonite (Fig. 5.3) is the best known example of this type. In this type, there is expanding space between the sheets. The cation exchange capacity is therefore greater in montmorillonite than kaolinite. The plasticity of montmorillonite is also higher because water can enter between the sheets.

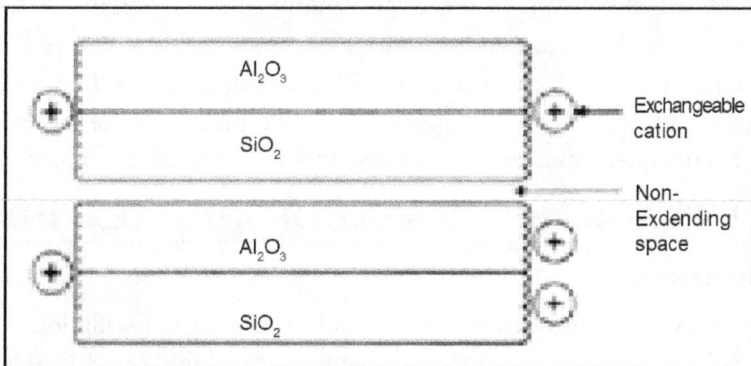

Fig. 3. Diagram of Kaollnite clay crystal

In three-layer type, there is another group called hydrous mica. Illite is the most important example of this group. Illite has similar structure as montmorillonite (2 :1 lattic structure). The structure is non- expanding type. Illite is in between Kaolinite and montmorillonite type with regards to soil properties.

c. Chlorites (2:1:1 types minerals). The typi-cal crystal unit is composed of one 2 : 1 unit (like mica or montmorillonite), and one octahedral unit, brucite layer. Magnesium domi-nates the octahedral sheet in the 2:1 units.

Chlorites are basi-cally iron-magne-sium silicates with some aluminium present. In a typical chlorite clay crystal, 2:1-type layers, alter-nate with a magne-sium-dominated trioctahedral sheet, giving a 2:1:1 ratio (Fig. 4). This mineral is sometimes called 2-2 type mineral.

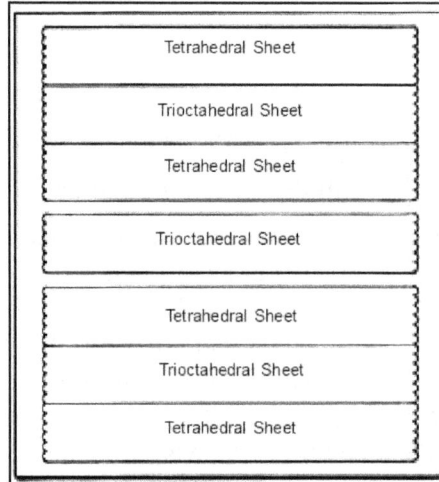

Fig. 4 Diagram of a Chlorites

ORGANIC COLLOIDS

Organic colloids are chiefly due to presence of humus in soil. Humus is the product of decomposition of plant and animal residues. Humus colloids are composed of carbon, hydrogen, oxygen and nitrogen, instead of silicon, aluminium and oxygen, as in clay colloids. Organic soil colloids have higher adsorptive properties for water and cations (Ca^{++}, K^+ etc.) and higher cation exchange capacity than colloidal clay (inorganic colloids).

PROPERTIES AND IMPORTANCE OF SOIL COLLOIDS

Brownian movement

Colloidal particles are found to be in continual motion. The oscillation is due to the collision of colloidal particles or molecules with those of the liquid in which they are suspended. This movement is mainly responsible for the coagulation or flocculation of colloidal particles. When the particles in suspension collide with each other and form a loose aggregate of floe.

Flocculation

The colloidal particles are coagulated by adding an oppositely charged ion. Formation of flocs is known as flocculation. If the cations are held close to the negatively charged particles, the negative charge would be neutralized and the colloidal particles flocculate and settle down. Na^+ (Sodium Cations) are highly hydrated and are monovalent; they are not so closely bound with the negatively charged immobile particles.

Thus, the particles continue to offer resistance to aggregation and do not flocculate. Ca^{++} (Calcium cations) are divalent and are not as easily displaced as sodium. Thus, calcium ions are able to neutralize the negative charge more efficiently and the colloidal system tends to flocculate. In a similar manner, trivalent ions like aluminium (Al^{+++}) are still more efficient in flocculation of colloids. Thus, Na-clay produces de-flocculation and Ca-clay encourages aggregation.

The phenomenon of flocculation plays an important part in the cultivation of soils. When clay particles are flocculated, soil develops small clods of a crumby nature. Such a soil allows free movement of air and water. If the particles are deflocculated, the aggregates get dispersed, the soil gets water-logged, the movement of air and water is impeded.

Electrical charge

Colloidal particles often have an electrical charge, some positive and some negative. When clay colloids suspended in water, they carry a negative electric charge. Collodial clay develops negative electric charge due to dissociation of hydroxyl groups attached to silicon in silica sheets of the clay mineral leaves residual oxygen (O^-) carrying a negative charge.

Adsorption

Colloidal particles possess the power of adsorbing gases, liquid and even solids from their suspension. The phenomenon of adsorption is confined to the surface of colloids particles. Larger the surface area (area per unit weight) greater the adsorption for water, nutrients etc. For example, take a cube of 1 cm edge.

The surface area exposed by this cube is equal to 6 sq. cm. When this cube is sub-divided into 8 cubes of edges 0.5 cm, the surface area exposed is 12 sq. cm. Thus, the smaller the size of the particle the greater the surface area exposed by them. The adsorption of ions is governed by the type and nature of ion and the type of colloidal particle.

In the case of cations, the higher the valence of the ion, the more strongly it is absorbed. Exchange or replacement of cation would be difficult from colloidal particle. That is why divalent ions like calcium and magnesium (Ca^{++} & Mg^{++}) are held more strongly than monovalent ion, sodium and potassium (Na^+ and K^+). Aluminium (Al^{+++}), a trivalent cation, is most easily adsorbed. Hydrogen ions (H^+) behave a polyvalent ions so are adsorbed more strongly than even Ca^{++}.

Adsorption of anions ($H_2PO_4^-$, HPO_4^- etc.) increases with the lowering or increasing of pH. The adsorption of phosphate ions is the lowest when the medium is neutral; it increases when the pH either falls or rises, due to fixation by iron and aluminium hydroxides in acid range and by calcium in alkaline range.

Among the clay minerals, kaolinitic clay has a greater anion adsorbing capacity than montmorillonic or illitic clay. The property of adsorption plays an important role in soil fertility. Due to this property soil is able to hold water and nutrients and keep them available to plant.

Non-permeability

Colloids, are unable to pass through a semipermeable membrane. The membrane allows the passage of water and of the dissolved substance through its pores, but retains the colloidal particles.

Cohesion and adhesion:

Unlike sand, clay particles possess the properties of cohesion and adhesion. While forming aggregates, the colloidal clay particles unite with each other by virtue of the property of cohesion. Clay particles envelop sand particles under the force of adhesion. The force of cohesion and adhesion are developed in the presence of water.

When colloidal substances are wetted, water first adheres to the particles and then brings about cohesion between two or more adjacent colloidal particles. Soil when dried, the particles remain united because of the force of molecular cohesion. These two forces help in the retention of water in the soil and thus used by plants and microorganism.

Swelling

A soil colloid when brought in contact with water they imbibe a certain quantity of water and swell and increase in volume.

Plasticity

Soil colloidal particles may present in gel condition posses the property of plasticity. Due to this property clay-colloids can be moulded in any shape.

GLOSSARY

Actinomycetes: A large group of rod-shaped or filamentous bacterium that includes some that cause diseases and some that are the sources of antibiotics. These soil microorganisms generally resemble fungi and have branched mycelium.

Active Carbon (AC): The portion of total soil organic carbon (matter) that is relatively easily metabolized or utilized by microorganisms. The half-life of active carbon ranges from a few days to a few years. Active carbon would include simple polysaccharides and glucose equivalent reduced sugars, amino acids and proteins, soluble and extractable carbon, and microbial biomass carbon, etc.

Aerobic: Aerobic means in the presence of oxygen or growing in the presence of oxygen. Aerobic soils have plenty of oxygenated air to carry out oxidative reactions, such as soil organic matter decomposition and nutrient cycling.

Aggregates: Primary soil particles (sand, silt, and clay) held together in a single mass or cluster, such as a crumb, block, prism or clod using organic matter as cementing material. Soil aggregates are usually greater than ten millimeters in diameter and formed by natural forces (such as alternate wetting-drying) and organic substances derived from root exudates, roots, soil animals and microbial by-products which cement primary particles into smaller aggregates or smaller aggregates into larger particles, such as macroaggregates.

Air Quality: Defined as a measure of the amount or concentration of pollutants emitted into the atmosphere and the dispersion potential of an area to dilute those pollutants. The common air pollutants are ground-level ozone, carbon monoxide, sulfur dioxide, nitrogen oxide (e.g. N_2O), radon, and emitted heavy metal dusts. These pollutants can harm field crops, public health, birds and other animals, and the environment, and cause property damage. Carbon dioxide and N_2O are greenhouse gases that are normally present in the atmosphere, but at higher levels may contribute to global warming.

Ammonium nitrate: A dry granular material manufactured by reacting nitric acid with anhydrous ammonia. One-half of the nitrogen in ammonium nitrate is in the nitrate form, and one-half is in the ammonium form.

Anaerobic: Absence of oxygen or growing in the absence of oxygen. Soils that are heavy textured (clay), compacted, wet or flooded tend to be anaerobic because they have less oxygenated air to carry out oxidative reactions. Anaerobesis of soil is also responsible for widespread soil-borne diseases.

Anhydrous ammonia: A gaseous material that is compressed and stored as a liquid. At 60°F, a gallon of anhydrous ammonia weighs 5.15 pounds. Due to the fact anhydrous ammonia needs to go through the nitrification process, it is more resistant to losses from the soil by leaching or denitrification because it is converted by bacterial action to the nitrate form more slowly than are other nitrogen sources.

Aqua ammonia: Anyhdrous ammonia dissolved in water. It is a low pressure solution and contains free ammonia, the amount of which increases as air temperature increases.

Available water: (Capacity) The amount of water released between *in situ* field capacity and the permanent wilting point (usually estimated by water content at soil matric potential of –1.5 Mpa). Available water is not the portion of water that can be absorbed by plant roots, which is plant specific.

Bacteria: A large group of single-celled microorganisms lacking chlorophyll and are prokaryotic (lacking a nucleus). Typically a few micrometers in length, bacteria have a wide range of shapes, ranging from spheres to rods and spirals. Bacteria are important for functioning of biochemical properties and/or processes. Bacteria are active in the soil for decomposing organic matter, recycling nutrients, and detoxification of contaminants. Bacteria dominate in disturbed (e.g. conventionally tilled) soils because they are generalist feeders that prefer aerobic (oxygenated) conditions, and survive in small soil pore spaces (micropores).

Basal Respiration (BR): Respiration, or oxygen used, by soil microbes to decompose organic matter or any crop residues as a food and energy source and released as carbon dioxide into the soil atmosphere. Soil respiration is a direct and sensitive assessment of soil antecedent biological activity. A high BR generally indicates higher soil microbial activity, however, not biological efficiency.

Buffer pH: A test used to estimate lime requirements in soil, which means the amount of agricultural lime required to neutralize the H+ from the soil exchange sites and in the soil solution.

Bulk Density: The mass (weight) of unit soil divided by the total volume occupied. An ideal soil has a bulk density of about 1.25 g cm^{-3}. Bulk density is the total soil porosity. Compacted soils have a bulk density of 1.5 g cm^{-3} or higher.

Carbon Sequestration Index (C_{index}): A new term explains the simple and integrated index that identifies the rate of carbon accumulation in the soil. This index can be used as an early and sensitive indicator of carbon or soil organic matter accumulation and could help explain the complex nature of global carbon cycling.

Carbon-Nitrogen Ratio (C/N): Ratio of the mass (weight) of organic carbon to the mass of total nitrogen in the soil, plants, or any other organic compounds. Soil C:N is generally 10:1 to carry out ecological functions.

Carbon-nitrogen ratio: The ratio of the weight of carbon to the weight of nitrogen in a plant material, soil or organism. For example, C/N for sawdust is about 500/1. (Sawdust contains about 500 pounds of carbon for every one pound of nitrogen.)

Catalyst: Chemical substances generally produced by microorganisms or plants that promote biochemical reactions by lowering the energy needed during a reaction. It is not changed or consumed during the reaction. Catalysts are important in the soil in speeding up biochemical reactions and are usually only present in minute amounts.

Cation exchange capacity: The ability of a soil to hold certain elements which have a positive charge; mainly a function of clay content and organic matter. Some plant nutrients, such as potassium and caldium are cations, so a soil with higher CEC will generally be more fertile because of its greater ability to hold these nutrients. Measured in milliequivalents/100 grams.

Cellulose: The greatest amount of carbon in a plant in the form of carbohydrates is cellulose, which gives plants structural rigidity, and allows plants to grow erect. As a plant matures, the cellulose content of the plant generally increases.

Clay: As a soil separate, clay refers to mineral soil particles which are less than 0.02 millimeters in diameter. As a soil textural class, clay refers to soil material that is 40 percent or more clay, less than 45 percent sand, and less than 40 percent silt.

Climate: Precipitation, temperature and other aspects of climate affect how specific soils are formed.

Conventional Tillage (CT): The combined secondary and primary tillage operations performed in preparing a seedbed in a given geographical location. Examples of primary tillage include moldboard plowing, chisel plowing, and sub-soiling, while finishing tools like a disk, field cultivator, or cultipacker would be considered secondary tillage. Conventional tillage means that the soil is physically turned over (or loosened), and oxygenate. This process accelerates breakdown of crop residues and native soil organic matter and temporarily increases microbial respiration. Soil organic matter decomposition releases nutrients, especially N, but it also destroys soil structure and macropores leading to soil compaction and decreased soil organic matter levels.

Decomposition: Biochemical breakdown of plant or animal residues into simpler compounds, often accomplished by soil food webs especially microorganisms.

Deductive Soil Quality: A measure of soil quality, based on crop production from the soil, such as grain yield, hay, fruits, and other outputs.

Denitrification: Reduction of nitrogen (e.g. nitrate or nitrite) to molecular nitrogen or nitric oxide by microbial activity or biochemical activity. Typically, it occurs under anaerobic (lack of oxygen) conditions when soil is wet and compacted. The result is a loss of nitrogen to the atmosphere.

Ecosystem Services: The resources and processes supplied by natural ecosystems that benefit humans. Collectively, these benefits are known as *ecosystem services* and include products like clean drinking water and processes such as the decomposition of wastes. Ecosystem services have been grouped into four broad categories: *provisioning*, such as the production of food and availability of water; *regulating*, such as the control of climate and diseases; *supporting*, such as nutrient cycles and crop pollination; and *cultural or social*, such as norms, ideas, and spiritual and recreational benefits.

Elemental sulfur: The most concentrated form of sulfur. Elemental sulfur must be oxidized to the sulfate form before plants can use it. It must be finely ground to particle sizes of 80-100 mesh to be oxidized and effective during the same growing season.

Eutrophication: Enrichment of natural fresh waters with excess nutrients (especially phosphorus) that lead to algal blooms and subsequent oxygen deficiencies when the algae die and start to decompose. However, both nitrogen and phosphorus are responsible for sea water eutrophication, such as in the Gulf of Mexico.

Field capacity: Soil water content after the soil has been saturated and allowed to drain freely for 24 to 48 hours. Free drainage occurs because of the force of gravity pulling on the water. When water stops draining, we know that the remaining water is held in the soil with a force greater than that of gravity.

Fulvic Acid (FA): Light yellow colored material that remains in solution after separating humic acid from humus by acidification. Fulvic acid is less dense, more alipathic, and has a low molecular weight compared to humin and humic acid but has a high nitrogen content that is more available to plants. It is soluble in any pH levels.

Fungal Hyphae: Long and often branched filaments associated with fungus, somewhat like a spider web. Fungal hyphae are important to extract nutrients, especially micronutrients, to support growth in poor quality soil.

Fungus: Microorganisms that contain a rigid cell wall that are either single-celled or multicellular organisms without chlorophyll that reproduces by spores and lives by absorbing nutrients from organic matter. Fungi include mildews, molds, mushrooms, rusts, smuts, and yeasts. Fungi survive better under no-till or undisturbed ecosystem and slightly acid conditions. They do not perform well in plowed fields due to their extensive hyphae (spider web like) networks affected by tillage operations.

Glomalin: A glycoprotein (sugar-protein complex) produced by mycorrhizal fungus (e.g. *Glomus* spp.). Glomalin binds soil particles together to form aggregates to promote soil aggregate stability and are important in the storage of soil carbon.

Grassland: Land on which the vegetation is dominated by grasses, grasslike plants, and/ or forbs. Herbaceous vegetation provides at least 80% of the canopy cover.

Hemicelluloses: Consists of polymers that branched, the second most common carbohydrate (up to 30%) in plant residues. Hemicelluloses generally contain of several types and forms due to condensation of sugars. Pectin is an example of hemicelluloses.

Horizons: The layers of soil, usually three, which make up the soil profile. Soil horizons differ in color, texture, structure and organic matter content. The A Horizon is the upper surface or topsoil and usually has the highest organic matter content; the B Horizon is the subsoil; and the C Horizon is the parent material. A given soil may have one or all three horizons.

Humic Acid (HA): Dark colored organic material extracted from soil by various reagents (such as dilute alkali) and is precipitated by acid. Humic acid has a relatively high molecular weight and is dense, aromatic, and dark brown to black in color. It is relatively stable compared to fulvic acid and essential for soil cation exchange capacity and formation of soil structures.

Humic Substances (HS): Series of relatively high molecular weight organic substances which are brown to black in color. Make up 60 to 80% of soil organic matter and are generally partial decomposable to resistant to microbial decomposition. They have a hydrocarbon structure composed of C-C, C-N, C=C, and C-O-C bonds.

Humification Degree (HD): The ratio of humic acid (HA) and fulvic acid (FA) to total humified organic matter, HD = (Humic acid + Fulvic acid)/ (Total extracted C)*100. The greater the HD, the higher the soil organic matter polymerization and condensation. In other words, high DH means the organic matter is more resistant to decomposition.

Humification Index (HI): A ratio of non-humified carbon or glucose carbon, divided by total humified organic carbon, HI = (Non-humic C)/ (Humic acid + Fulvic acid)*100. Higher HI indicates less humification of soil organic matter.

Humification Ratio (HR): The ratio of humic acid plus fulvic acid divided by total carbon, HR = (Humic acid + Fulvic acid)/ (Total organic C)*100. Low HR indicates less humification of soil organic matter.

Humification: Process by which the carbon of organic substances is transformed and resynthesized to humic substances through biochemical and physio-chemical processes.

Humin (HN): A fraction of soil organic matter that cannot be extracted from soil with dilute alkali or acid. Humin is a very stable fraction of soil organic matter due to its strong association or bonding with clay minerals and metal ions.

Humus: An organic material in soil which is a product of plant and animal remains that have decomposed and then synthesized into something new.

Humus: Total of the organic compounds in the soil excluding undecayed plant and animal tissues. Humus equals fulvic acid plus humin, plus humic acid. The term is often used to describe soil organic matter. Humus is generally dark in color.

Hypoxia: Insufficient oxygen in an environment to support aerobic life. Usually occurs in estuarine and sea water (including Gulf of Mexico) from an excess of nitrogen and phosphorus which promotes the growth of algae. When the algae die, the decomposition uses so much oxygen other life is harmed.

Inductive Soil Quality: A measure of soil quality based on soil characteristics and/ or processes that regulate the functional behavior of the soil, such as hydrologic properties, microbial activity, emission of gases, organic matter, and nutrient availability.

infiltration: Water movement in the soil. Pore space in soil is the conduit that allows water to infiltrate and percolate (downward movement of water through the soil).

Labile: A substance that is readily transformed or used by microorganisms. Labile organic matter is often associated with active organic carbon to regulate soil quality.

Legume: Legumes are plants associated with nitrogen fixing organisms and include peas, beans, peanuts, clovers, alfalfa, lespedezas, vetches, and kudzu.

Lignin: Lignin is the third most common component of plant residue. It is prominent in woody or mature tissues and increases as a plant ages. Lignin is a very complex energy-rich molecule and varies greatly in structure, making it hard to decompose. Lignin helps cement cell walls together to provide structural support for plants.

lime: A naturally occurring material composed of carbonates of calcium and magnesium.

Macroaggregates: Soil aggregates greater than 250 micrometers in size consisting of microaggregates cemented together by organic matter, microbial polysaccharides, fungal hyphae, earthworm excretions, and plant roots. Macroaggregates are typically found in undisturbed soils such as continuous no-till with cover crops.

Macropores: Larger soil pores (greater than 60 micrometers) from which water drains readily by gravity. Macropores are important for soil aeration and good drainage.

mechanical analysis: The laboratory procedure used to identify soil separates.

Mesofauna: Animal life of medium size (between 2 and 0.2 millimeters in diameter). Examples include centipedes, nematodes, and mites.

Metabolic Quotient (qR): A ratio of total microbial biomass carbon over total organic carbon. A high qR is an indication of an enlarging biological carbon pool in total organic carbon. It typically ranges from 0.5 to 5% in soil.

Microaggregates: Soil aggregates less than 250 micrometers in size consisting of primary particles, plant roots, and humin cemented together. Microaggregates are more typically found in disturbed or cultivated soils. Multiple microaggregates may form larger macroaggregates through microbial activity, plant root exudates and actions, fungal hyphae, and earthworm casts.

Microbial Biomass: Includes the smallest living organisms such as bacteria, fungus, protozoa, algae, actinomycetes, nematodes, and nonliving organisms: prion and viruses. Microbial biomass denotes a small portion (less than 5%) of soil organic carbon. The main function of the microbial biomass is to act as bio-catalyst for organic matter decomposition and mineralization, soil fertility, and humus formation and soil aggregation.

Microfauna: Animal life too small to be clearly seen without a microscope; includes protozoa and nematodes.

Microflora: Plants too small to be clearly seen without using a microscope; includes actinomycetes, algae, bacteria, and fungus.

Micronutrients: Elements needed in very small amounts for plant growth. Micronutrients include zinc, iron, chlorine, copper, manganese, boron and molybdenum.

Micropores: Smaller soil pores (less than 60 micrometers) generally found within soil aggregates. Water does not drain freely in micropores.

Mineralization: Conversion of an organic compound to a transitional or inorganic form by microorganisms. As a result, nutrients are generally more available and absorbed by plant roots.

Mycelium: A string-like mass of individual fungal or actinomycetes hyphae. Mycelium is the vegetative part of a fungus, consisting of a mass of branching, threadlike hyphae.

Mycorrhizae: Literally means "fungus root" and is a symbiotic (mutually beneficial) relationship between fungus and plant roots. The fungus supplies water and nutrients to the plant roots while the plant supplies carbohydrates. Plant roots typically can explore no more than 1% of the soil volume but with mycorrhizal fungus (which attach themselves to the plant root cell walls) association, approximately 20% of the soil volume may be explored. Over 80% of plants have a mycorrhizal association but these fungus populations are reduced by conventional tillage and high fertilizer applications of nitrogen and phosphorus.

Nematode: An unsegmented microscopic roundworm located near plant roots. Nematodes feed on plants, animals, bacteria, and fungi. Typically found in higher concentrations in no-till and undisturbed soils. They are the most populous animals on the planet. In fact, if we counted all the animals on earth, four out of every five would be a nematode.

Nitrogen: Element needed in large amounts for plant development; found naturally and in applied fertilizers.

Non-Humified Carbon: Glucose equivalent or total reducing sugars or amino sugars. This is the most labile fraction of carbon in humus to decompose.

No-Till: A system where the crop is planted directly into a seedbed without tilling or disturbing the entire soil surface. The only soil disturbance is for placement of seed and fertilizer. This system is also called zero-till or direct seeding. No-till farming, by definition, means that the soil has not been disturbed since the prior harvest of a crop. Continuous no-till means the soil has not been disturbed for several years.

organic matter: Material that contains carbon and is found in the soil. Most soil organic matter comes from previously living organisms. Temperature and moisture are the two main factors affecting its development.

parent material: Rock or minerals which are weathered to form smaller particles of a soil. Parent material is one of the five factors contributing to formation of a specific soil. In the Great Plains, much parent material is associated with ancient seas or glacier deposits.

particle density: The weight of an individual soil particle per unit volume. Particle density is usually expressed in units of grams per cubic centimeter (g/cm^2). Bulk density considers both the solids and the pore space; whereas, particle density considers only the mineral solids.

Passive Carbon: Passive pools of carbon that have a slow turnover rate and are more resistant to decomposition and nutrient release. Example: more humified soil organic matter. This fraction provides substrate (food and energy) to microbes and acts as binding agents for soil aggregate formation.

percolation: The downward movement of water through the soil, made possible by pore space in the soil.

pH: The measurement of an aqueous solution's acidity and alkalinity; measured on a scale of 1 to 14. Pure water has a pH of 7.0 and is neutral. Different crops grow best at different pH levels; pH influences herbicide activity and nutrient uptake.

phosphorus: A key element in the complex nucleic acid structure of plants which regulates protein synthesis; important in cell division and development of new tissues. Next to nitrogen, the most limiting nutrient in Nebraska crop production; naturally found in sufficient amounts in many Nebraska soils.

pore space: Spaces in soil, between the mineral and organic matter, that are filled with water or air.

porosity: The volume of soil voids that can be filled by water and/or air; inversely related to bulk density. Porosity is also known as "pore space."

Porosity: Porosity is a measure of the void spaces in a soil, represented as the volume of voids divided by the total volume of soil. In an ideal soil, the total pore space should be about 50% (composed of air and water) while the solid phases (sand, silt, clay, and organic matter) make up the other 50% of soil volume.

Potassium: An essential plant nutrient needed in large amounts. Postassium is vital to plant nutrient absorption, respiration, transpiration and enzyme activity. The major portion of potassium is contained in minerals such as feldspar and mica, and clays such as montmorillonite, vermiculite and illite.

Potentially Mineralizable Carbon: Soil organic carbon that is mineralized during microbial decomposition of organic matter or under controlled incubation.

profile: The shape of a slope profile. Profile is the slope viewed in a vertical cross-section.

Protozoa: Any of a large group of one-celled organisms that move by flagella (flagellates), cilia (ciliates), or have flow (amoeba). Most species feed on bacteria, fungi, or dead microbial particles. Protozoa are more common in conventional tilled or disturbed soils where they outnumber the nematodes.

Respiration: The biochemical processes by which all living organisms derive energy from food or stored reserves by taking oxygen from the environment and giving off carbon dioxide.

Rhizobia: A nitrogen-fixing bacterium (genus *Rhizobium*) that is common in the soil, especially in the root nodules of leguminous plants.

Rhizosphere: A soil zone near the plant roots where microbes flourish in greater numbers and have more activity than in the bulk soil. The rhizosphere typically supports 1000 to 2000 times more microbes than the bulk soil without live roots. Roots give off many root exudates which supply food for the microbes and increases microbial activity.

Sand: Individual rock or mineral fragments in a soil that range from 0.05 to 2.0 millimeters in diameter. Most sand grains consist of quartz, but they can be of any mineral composition. Sand is also the textural class name of any soil that contains 85 percent or more sand and no more than 10 percent clay.

Silt: A soil inorganic separate in the range of 2 to 50 micrometers (or 0.002 to 0.05 mm.). Silt is smaller than sand but larger than clay.

soil: The top layer of the Earth's surface, consisting of four major components: air, water, organic matter and mineral matter. There are three categories of soil particles—sand, silt and clay—which are called "soil separates."

Soil age: Determined by the amount of weathering that has occurred; to what extent the parent material has been converted to distinct horizons or soil layers. Usually described as young, mature or old.

Soil Biological Quality: Biological properties that are associated with soil functionality including microbial biomass, nematodes, earthworms, biological activity, and enzymes.

Soil Carbon Sequestration: How carbon is stored in the soil; the process by which carbon dioxide from the atmosphere is converted to organic carbon by photosynthesis and the decomposition of that plant carbon into organic matter stored in the soil.

Soil Chemical Quality (SCQ): Chemical properties that are associated with soil quality including carbon, nitrogen, cation exchange capacity and base saturation, and salt content.

Soil classification: A specific soil is classified according to the number of horizons in its soil profile and the soil properties of each horizon. Well developed soils are old and contain all three master horizons (A, B, and C), as well as several subdivisions of the master horizons.

Soil development: Five factors influence the development of a specific soil: parent material, climate, living organisms, topography and time.

Soil Functions: Ecosystem functions are directly and indirectly associated with soil. Soils perform many functions, and healthy soil gives us clean air and water, bountiful crops and forests, productive rangeland, diverse wildlife, and beautiful landscapes.

Soil Health: Soil health is an assessment of ability of a soil to meet its range of ecosystem functions as appropriate to its environment. Also refers to the condition of the soil, including its ecosystems (minerals, nutrients, and microbial activity), pH, and structure. The Soil Science Society defines soil health as the capacity of a specific kind of soil to function, within natural or managed ecosystem boundaries, to sustain plant and animal productivity, maintain or enhance water and air quality, and support human health and habitation.

Soil Organic Carbon: Is related directly to soil organic matter. Soil organic matter consists of 50 to 58 percent soil organic carbon. A simple unit of 1.724 is used to convert total carbon into soil organic matter.

Soil Organic Matter: Classified into two major groups composing a humic and non-humic substances. Soil organic matter is thermodynamically unstable and is part of the natural balance between production, decomposition, transformation, and resynthesis of various organic substances. The humified fraction is composed of humic, fulvic, and humin and is the most stable. The non-humic portion is the relatively unstable and most labile fraction and is most easily decomposed.

Soil Physical Quality: All the physical properties and/or processes associated with soil including bulk density, porosity, infiltration, and micro- and macroaggregates.

Soil Quality Index: Integrated measure of soil quality by transforming and combining selected core biological, chemical, and physical properties into a single index to evaluate a soil's functional capability.

Soil Quality Indicators: A measure of a soil's functional state. Scientists use soil quality indicators to evaluate how well soil functions since soil function often cannot be directly measured. Measuring soil quality is an exercise in identifying soil properties that are responsive to management practices, affect or correlate with environmental outcomes, and are capable of being precisely measured within certain technical and economic constraints. Soil quality indicators may be qualitative (e.g. drainage is fast) or quantitative (infiltration = 2.5 in/hr). Ideal soil quality indicators should (1) correlate well with ecosystem processes; (2) integrate soil physical, chemical, and biological properties and processes; (3) be accessible to many users; and (4) be sensitive to management and natural processes.

Soil Quality: The capacity of a soil to function within natural or managed ecosystem boundaries, to sustain plant and animal productivity, maintain or enhance water and air quality, and support human health and habitation. Soil quality is analogous to soil health. Soil quality is the result of combined activities of biological, chemical, and physical properties as a reaction of management operations. Crop rotations, no-till, and cover crops improve soil quality.

Soil sample: A collection of individual cores from a known area.

Soil separates: Categories of soil particles—sand, silt and clay—divided by particle size. The proportion of different soil separates in a field defines its soil texture.

Soil series: A unit of soil classification determined by studying horizon characteristics, such as: number of horizons, color, thickness, texture, erosion phase, slope, organic content and depth to hardpan. All soils given the same soil series name possess the same characteristics across the landscape.

Soil Structure: Combination or arrangement of soil primary particles into secondary units or peds (composed of macroaggregates and microaggregates). The secondary units are classified on the basis of size (microaggregates are the smallest and macroaggregates are the largest) and shape. Soils with good structural stability typically have more macroaggregates and macropores while soils with poor structural stability have more microaggregates and micropores. Compacted soils have poor structure and more microaggregates and micropores.

Soil test: Chemical analysis of soil samples to assess soil nutrient levels and determine how fertilizer use can be improved.

Soil Texture: Relative portion of sand, silt, and clay in a given amount of soil. Texture is important to evaluate soil quality associated with fertility and crop productivity, compaction, hydrological properties, construction, and biological activity.

Specific Maintenance of Respiration (qCO^2): It is expressed as basal respiration per unit of total microbial biomass. It is an indicator of soil ecosystem disturbance. A low qCO^2 indicates low disturbance and stable no-till situation or a more mature ecosystems; a high qCO^2 indicates highly disturbed ecosystems as found under conventional tillage.

structure: The combination or arrangement of soil particles that forms peds or aggregates.

terraces: Physical earthen barriers to runoff and can be very effective in reducing P losses to erosion and runoff.

Total Carbon: The total of all forms of carbon in the soil. If the pH of the soil is 6.5 or higher, all the total carbon is considered to be organic carbon. Total carbon (TC) minus active carbon (AC) equals passive carbon (PC). Total carbon is composed of microbial carbon, particulate organic carbon, fulvic acid, humic acid, humin, non-humified carbon (glucose), and calcium and magnesium carbonates.

Total Nitrogen: Total of all forms of nitrogen in the soil. Total nitrogen also includes the ammonia that is non-mobile and the nitrate form which is more mobile.

Total Phosphorus (TP): Total of all forms of phosphorus in the soil including organic (inositol, phospholipids, etc.) and inorganic (phosphates) phosphorus.

Urea: A dry nitrogen material produced by reacting ammonia with carbon dioxide. Urea contains the highest percentage of nitrogen of the commonly used dry fertilizers and is rapidly replacing ammonium nitrate. When surface applied, urea is the most rapidly volatilized of the dry nitrogen materials.

Virus: A large group of submicroscopic infective agents that typically contain a protein coat surrounding a nucleic acid core. Typically considered nonliving and are capable of growth only in a living cell.

Water Quality: Defined as the biological, chemical, and physical conditions of water; a measure of the ability of water to support beneficial uses. The composition of water as affected by natural processes and human activities depending on the water's chemical, biological, physical, and radiological condition.

Water Stable Aggregates: A soil aggregate that is stable to the action of water such as falling raindrops or agitation, as in wet-sieving analysis. Water stable aggregates improve soil quality.

water-holding capacity: The ability of a soil to hold water; varies by soil texture. Medium textured soils (fine sandy loam, silt loam and silty clay loam) have the highest water-holding capacity.

Weathering: The means by which soil, rocks and minerals are changed by physical and chemical processes into other soil components. Weathering is an integral part of soil development.

BIBLIOGRAPHY

Abeele, W.V., Consolidation and Shear Failure Leading to Subsidence and Settlement Part I, November 1985, Los Alamos National Laboratory, Los Alamos., New Mexico 87545, 1985.

Agbede, O.A., Jatau, N.D., Oluokun, G.O. and Akinniyi, B.D., 2015, Geotechnical investigation into the causes of cracks in building: A case study of Dr. Egbogha Building, University of Ibadan, Nigeria., IJESI, 4 (11), 18-22.

Akayuli, C., Ofosu, B., Nyako, S.O. and Opuni, K.O., 2013, The influence of observed clay content on shear strength and compressibility of residual sandy soils., Int J Eng Res Appl., 3 (4), Jul-Aug, 2538-2542.

Alberty, C.A., H.M. Pellett, and D.H. Taylor. 1984. Characterization of soil compaction at construction sites and woody plant response. J. Environ. Hortic. 2(2):48-53.

Apparao, K.V.S. and V.C.S. Rao, Soil Testing Laboratory Manual and Question Bank, Universal Science Press, New Delhi, 1995.

Arora, K.R., Soil Mechanics and Foundation Engineering (Geotechnical Engineering), Standard Publishers Distributors, Delhi, 2008.

Assouline, S., K. Narkis, K, S.W. Tyler, I. Lunati, M.B. Parlange, and J.S. Selker. 2010. On the Diurnal Soil Water Content Dynamics during Evaporation using Dielectric Methods Vadose Zone J., 9 (3): 709-718.

Assouline, S., S.W. Tyler, J.S. Selker, I. Lunati, C.W. Higgins, M.B. Parlange. 2013. Evaporation from a shallow water table: Diurnal dynamics of water and heat at the surface of drying sand. Water Resour. Res 49(7):1944-7973. DOI 10.1002/wrcr.20293.

Blake, G.R., and K.H. Hartge. 1986. *Bulk density, radiation methods*. In Klute, A. (Ed.). Methods of Soil Analysis, Part 1. Physical and Mineralogical Methods. Agronomy Monographs, No. 9 (2nd ed.). American Society of Agronomy, Madison, WI.

Bowles, J. E., Engineering Properties of Soils and their Measurements, 4th edition, McGraw Hill Education (India) Private Limited, New Delhi, 2012.

Brady, N.C. 1974. The Nature and Property of Soils. 8th Ed. Macmillan Publishing Company, Inc., New York, N.Y.

British Standards Institution, BS 1377-7:1990 – Methods of test for soils for civil engineering purposes. Shear strength tests (total stress)

British Standards Institution, BS 1377-8:1990 – Methods of test for soils for civil engineering purposes. Shear strength tests (effective stress)

British Standards Institution, BS 8002:2015 – Code of Practice for Earth Retaining Structures

British Standards Institution, BS EN 1997-2:2007 – Eurocode 7 – Geotechnical design Part 2: Ground investigation and testing

Brown, R.A. and Hunt, W.F. (2010). "Impacts of construction activity on bioretention performance." Journal of Hydrologic Engineering. 15(6), 386-394.

Brutsaert, W. 1982. Evaporation into the Atmosphere. P. Reidel Publishing Company. Dordrecht, Holland.

Campbell, James. *Introduction to Remote Sensing.* New York: Guilford Press, 1987.

Carslaw, H.S., and J.C. Jaeger. 1959. The Conduction of Heat in Solids. Oxford University Press. London.

Chaplin, Jonathan, Min Min, and Reid Pulley. 2008. Compaction Remediation for Construction Sites. Final Report. Department of Bioproducts and Biosystems Engineering, University of Minnesota, St. Paul, Minnesota.

Charles Wortmann, Martha Mamo, and Charles Shapiro, Lime Use for Soil Acidity Management, University of Nebraska Extension NebGuide G03-1504-A.

Charles Wortmann, Martha Mamo, and Charles Shapiro, Management Strategies to Reduce the Rate of Soil Acidification, University of Nebraska Extension NebGuide G03-1503-A.

Chen, B.S, and Jensen, R.E., 2013, Case Studies of Dewatering and Foundation Design: Retail Warehouses in Taiwan., Seventh Internal Conference on Case Histories in Geotechnical Engineering, Chigo, Paper No. 3.03c, 1-10.

Cogger, C. *Potential Compost Benefits for Restoration of Soils Disturbed by Urban Development.* Compost Science & Utilization 13.4 (2005): 243-251.

Conforti, M., Muto, F., Rago, V. and Critelli S. (2014). Landslide inventory map of north-eastern Calabria (South Italy), Journal of Maps, 10:1, 90-102, DOI: 10.1080/17445647.2013.852142

Corominas, J., Mavrouli, O. and Roger R.C. (2017). Rockfall Occurrence and Fragmentation. 75-97. 10.1007/978-3-319-59469-9_4.

Dafalla, M.A., 2013, Effects of clay and moisture content on direct shear tests for clay-sand mixtures, Adv. Mater. Sci. Eng., Volume 2013, http://dx.doi.org/10.1155/2013/562726, 1-8.

Durgunoglu, H.T., Varaksin, S., Briet, S. and Karadayilar, T. (2003) A case study on soil improvement with heavy dynamic compaction, Proc. XIII ECSMGE, vol. 1, 651-656.

El-Maksoud, M.A.F., 2006, Laboratory determining of soil strength parameters in calcareous soils and their effect on chiseling draft prediction, Proc. Energy Efficiency and Agricultural Engineering, Int. Conf., Rousse, Bulgaria.

Ersoy, H., Karsli, M.B., Cellek, S., Kul, B., Baykan, I. and Parsons, R.L., 2013, Estimation of the soil strength parameters in Tertiary volcanic regolith (NE Turkey) using analytical hierarchy process., J. Earth Syst. Sci., 122 (6), December, 1545–1555.

Europa Orbiter Project. Jet Propulsion Laboratory, National Aeronautics and Space Administration. <http://www.jpl.nasa.gov/europaorbiter/>.

Ferguson, R.B., C.A. Shapiro, and G.W. Hergert. 1994. Fertilizer Nitrogen Best Management Practices. NebGuide G94-1178A. University of Nebraska, Cooperative Extension Service, Lincoln, NE.

GEER (2020). The September 18-20 2020 Medicane Ianos Impact on Greece - Phase I Reconnaissance Report. GEER-068, https://doi.org/10.18118/G6MT1T

GIS Data for Water Resources. U.S. Geological Survey. <http://water.usgs.gov/GIS/>.

GIS.com. <http://www.gis.com/>.

Gogoi, J.C. and Laskar, A.A., 2015, Mechanical compaction - a simple ground improvement technique: a case study., Discovery, 40(185), 377-383.

Gregory, J.H., Dukes, M.D., Jones, P.H., Miller, G.L., 2006. Effect of urban soil compaction on infiltration rate. Journal of Soil and Water Conservation 61, 117-124.

Hamilton, G.W., Waddington, D.V., 1999. *Infiltration rates on residential lawns in central Pennsylvania.* Journal of Soil and Water Conservation 54, 564-568.

Hanks, D. and A. Lewandowski, 2003. Protecting Urban Soil Quality: Examples for Landscape Codes and Specifications. USDA Natural Resources Conservation Services.

Hanks, R.J., 1992. Applied Soil Physics. 2nd ed., Springer Verlag. New York. NY.

Harper, J.F. 1984. Uptake of Organic Nitrogen Forms by Roots and Leaves. In R.D. Hauck (ed.) Nitrogen in Crop Production. American Society of Agronomy, Madison, WI.

Hase, W.J., et al., 1957. Nitrogen and Carbon Changes in Great Plains Soils as Influenced by Cropping and Soil Treatments, Tech. Bull. 1167 (Washington, D.C.: U.S. Dept. Agric.)

Hazen, A. (1911). Discussion of "Dames on sand formation," by A.C. Koenig. Transactions of the American Society of Civil Engineers, 73, 199-203.

Head, K. H., Manual of Soil Laboratory Testing, John Wiley & Sons, Inc., New York, 1982.

Highland, L. and Bobrowsky, P. (2018). TXT-tool 0.001-2.1 Landslide Types: Descriptions, Illustrations and Photos. 10.1007/978-3-319-57774-6_1.

Hillel, D. 1980. Fundamentals of Soil Physics. Academic Press, San Diego, CA.

Hiraiwa, Y. and T. Kasubuchi. 2000. Temperature dependence of thermal conductivity of soil over a wide range of temperature (5-75!). European Journal of Soil Science. vol. 51(2): 211-218.

Hofmann-Wellenhof, B., H. Lichtenegger, and J. Collins. *Global Positioning System: Theory and Practice.* New York: Springer-Verlag Wien, 2001.

Holz, Robert K. *The Surveillant Science: Remote Sensing of the Environment.* New York: John Wiley & Sons, 1985.

Indiana Department of Transportation. 2003. Dynamic Cone Penetration Test (DCPT) for Subgrade Assessment. Publication No. FHWA/IN/JTRP-2002/30, SPR-2362.

IS: 2720 – Part 5, 1970, Determination of liquid and plastic limits, BIS, New Delhi.

IS: 2720 – Part 6, 1972, Determination of shrinkage factors, BIS, New Delhi.

IS:2720 – Part 14, 1983, Determination of density index (relative density) of cohesionless soils, BIS, New Delhi.

Jain, V.K., Dixit, M. and Chitra, R., 2015, Correlation of plasticity index and compression index of soil., IJIET., 5(3), June, 263-270.

Jury, W.A., W.R., Gardner, and W.H. Gardner. 1991. Soil Physics. John Wiley and Sons, New York, NY.

Kaniraj, S.R., Design Aids in Soil Mechanics and Foundation Engineering, McGraw Hill Education (India) Private Limited, New Delhi, 1988.

Karmi, M.V., Mehrdad, M.A. and Eslami A., 2006, Simultaneous effect of height of shear strength parameters on optimization of embankment dams-two case studies., Dams and Reservoirs, Societies and Environment in the 21st Century, L. Berga, ýJ.M. Buil, ýE. Bofill, J.C.De Cea, J.A. Garcia Perez, G. Manueco, J. Polimon, A. Soriano & J. Yague (eds), Taylor & Francis Group, London, pp. 945-950, https://books.google.co.in/books?isbn=1134138504.

Karsten, T.K., Gau, C. and Tiedemann, J., 2006, Shear strength parameters from direct shear tests - influencing factors and their significance., IAEG2006 Paper number 484, The Geological Society of London, 1-12.

Katupitiya, A. 1995. Long-term Tillage Effects on Nitrate Movement and Accumulation and Denitrification in the Root and Intermediate Vadose Zones. Ph.D. Dissertation, University of Nebraska.

Kees, Gary. 2008. Using Subsoiling To Reduce Soil Compaction. U.S. Forest Service Technology & Development Publication 3400 Forest Health Protection 0834-2828-MTDC. Downloaded from http://www.fs.fed.us/t-d/pubs/htmlpubs/htm08342828/index.htm July 15, 2013

Kissel, D.E. 1989. Management of Urea Fertilizers. North Central Regional Publication No. 326.

Koçak, A. and Köksal, K., 2010, An example for determining the cause of damage in historical buildings: Little Hagia Sophia (Church of St. Sergius and Bacchus) – Istanbul, Turkey., Eng. Fail. Anal., 17, 926–937.

Laskar, A. and Pal, S.K., 2012, Geotechnical characteristics of two different soils and their mixture and relationships between parameters., EJGE, 17, 2821-2832.

Lichter, J.M., and L.R. Costello. 1994. *An evaluation of volume excavation and core sampling techniques for measuring soil bulk density.* J. Arboric. 20(3): 160-164.

Lo, C. P., and Albert K. W. Yeung. *Concepts and Techniques of Geographic Information Systems.* Upper Saddle River, NJ: Prentice Hall, 2002.

Look, B. (2007). Handbook of geotechnical investigation and design tables. London: Taylor & Francis. Ortiz et al. (1986).

Mallo, S.J. and Umbugadu, A.A., 2012, Geotechnical study of the properties of soils: a case study of Nassarawa – Eggon town and Environs, Northern Nigeria., CJEarthSci., 7 (1), 40 – 47.

Mars Exploration: Mars Express. Jet Propulsion Laboratory, National Aeronautics and Space Administration. <http://mars.jpl.nasa.gov/missions/future/express.html>

Mollahasani, A., Alavi, A.H., Gandomi, A.H. and Rashed, A., 2011, Nonlinear neural-based modeling of soil cohesion intercept., KSCE J CIV ENG., 15(5), 831-840.

Murthy, V.N.S (1996) A Text Book of Soil Mechanics and Foundation Engineering, UBS Publishers' Distributors Ltd. New Delhi, India.

Naik, S., Naik, N.P., Kandolkar, S.S. and Mandrekar, R.L., 2011, Settlement structure – A case study., Proc. Indian Geotechnical Conference, Kochi, 1031-1034.

Natural Resource Conservation Service. 1998. Soil Quality Test Kit Guide.

Natural Resource Conservation Service. 2012. National Engineering Handbook. Chapter 11, Cone Penetrometer. Part 631.

New York Department of Transportation. 2015. Geotechnical Test Method: Test Method For Earthwork Compaction Control By Nuclear Gauge. Geotechnical Engineering Bureau. GTM-10, Revision#5.

Ngah, S.A. and Nwankwoala, H.O., 2013, Evaluation of geotechnical properties of the sub-soil for shallow foundation design in Onne, Rivers State, Nigeria., The IJES., 2 (11), 08 – 16.

Nwankwoala, H.O. and Amadi, A.N., 2013, Geotechnical investigation of sub-soil and rock characteristics in parts of Shiroro-Muya-Chanchaga area of Niger State, Nigeria., IJEE., 6(1), 8 – 17.

Nwankwoala, H.O. and Warmate, T., 2014, Geotechnical assessment of foundation conditions of a site in Ubima, Ikwerre Local Government Area, Rivers State, Nigeria., IJERD, 9(8), 50 – 63.

Ocean County Soil Conservation District (OCSCD), Schnabel Engineering Associates, Inc., USDA Natural Resources Conservation Service. 2001. Impact of soil disturbance during construction on bulk density and infiltration in ocean county. New Jersey.

Oghenero, A.E., Akpokodje, E.G. and Tse, A.C., 2014, Geotechnical properties of subsurface soils in Warri, Western Niger Delta, Nigeria., Journal of Earth Sciences and Geotechnical Engineering., 4(1), 89 – 102.

Oke, S.A. and Amadi, A.N., 2008, An assessment of the geotechnical properties of the sub-soil of parts of Federal University of Technology, Minna, Gidan Kwano Campus, for foundation design and construction., J Sci Educ Technol., 1 (2), 87 – 102.

Olsen, H.J. 1990. *Construction of an electronic penetrometer for use in the field.* Computers and Electronics in Agriculture. 5:1:65-75.

Olson, N.C. J.S. Gulliver, J.L. Nieber, and M. Kayhanian. 2013. *Remediation to improve infiltration into compact soils.* Journal of Environmental Management. 117:85-95.

Olson, Nicholas Charles. 2010. Quantifying the Effectiveness of Soil Remediation Techniques in Compact Urban Soils. University Of Minnesota Master Of Science Thesis.

Owen, Gordon T. 1987. *Soil disturbance associated with deep subsoiling in compact soils.* Canadian Agricultural Engineering: 33-37.

Oyediran, A. and Durojaiye, H.F., 2011, Variability in the geotechnical properties of some residual clay soils from south western Nigeria., IJSER., 2 (9), 1-6.

Pennsylvania Department of Environmental Protection. 2006. Pennsylvania Stormwater Best Management Practices Manual. BMP 6.7.3: Soil Amendment & Restoration.

Pennsylvania Department of Environmental Protection. 2007. Pennsylvania Stormwater Best Management Practices Manual. Section 5.6.2. Minimize Soil Compaction in Disturbed Areas.

Peterson, T.A., T.M. Blackmer, D.D. Francis, and J.S. Schepers. 1993. Using a Chlorophyll Meter to Improve N Management. NebGuide G93-1171A. University of Nebraska, Cooperative Extension Service, Lincoln, NE.

Pitt, R., Chen, S., Clark, S.E., Swenson, J., and Ong, C.K. 2008. "Compaction's impact on urban storm-water infiltration." Journal of Irrigation and Drainage Engineering. 134(5), 652-658.

Poulos, S.J., Liquefaction Related Phenomena; In: Advanced Dam Engineering for Design Construction and Rehabilitation Van Nostrand Reinhold (ed.) Jansen R. B., pp. 292–320, 1989.

Prakash, S. and P.K. Jain, Engineering Soil Testing, Nem Chand & Bros, Roorkee, 2002.

Raj, P. P., Soil Mechanics and Foundation Engineering, Dorling Kindersley (India) Pvt. Ltd., New Delhi, 2012.

Randrup, T.B. 1993. *Soil Compaction and Plant Growth* (Jordkomprimering og Plantevaskst). The Royal Veterinary and Agricultural University. Danish Forest and Landscape Research Institute. Copenhagen, Denmark. 78 pp.

Randrup, T.B. 1997. *Soil Compaction on Construction Sites.* Journal of Arboriculture, 23:5:207-210.

Ranjan, G. and A.S.R. Rao, Basic and Applied Soil Mechanics, New Age International (P) Ltd., Publishers, New Delhi, 1991.

Ranjan, G. and Rao, A.S.R. (2000) Basic and Applied Soil Mechanics. New Age International Publisher, New Delhi, India.

Roy, S. and Dass, G., 2014, Statistical models for the prediction of shear strength parameters at Sirsa, India., I. Journal of Civil and Structural Engineering., 4(4), 483-498.

Roy, S., 2016, Assessment of soaked California Bearing Ratio value using geotechnical properties of soils., Resources and Environment., 6(4), 80-87.

Sabins, Floyd. *Remote Sensing: Principles and Interpretation.* New York: W. H. Freeman and Co., 1987.

Schueler, T. 2000. *The Compaction of Urban Soil: The Practice of Watershed Protection.* Center for Watershed Protection, Ellicott City, MD. Pages 210-214 downloaded from http://www.cwp.org/online-watershed-library/cat_view/63-research/75-monitoring

Schueler, T. R. *Can Urban Soil Compaction Be Reversed?* Technical Note 108# from Watershed Protection Techniques. 669-666 :(4)1.

Schueler, T.R., and H.K. Holland. 2000. *The Practice of Watershed Protection; Techniques for Protecting our Nation's Streams, Lakes, Rivers, and Estuaries.* Mahler Books, Austin ,TX. Published by the Center for Watershed Protection. 742 pp.

Schultz, Gert, and Edwin Edwin. *Remote Sensing in Hydrology and Water Management.* Berlin, Germany: Springer-Verlag, 2000.

Selbig, W.R., and N. Balster. 2010. Evaluation of turf-grass and prairie-vegetated rain gardens in a clay and sand soil: Madison, Wisconsin, water years 2004–08. U.S. Geological Survey, Scientific Investigations Report 2010–5077, 75 p.

Shanyoug, W., Chan, D., Lam, K.C., 2009, Experimental study of the fines content on dynamic compaction grouting in completely decomposed granite of Hong Kong., Constr Build Mater., 23, 1249 -1264.

Skempton, A.W., The Colloidal activity of clays; Proc. 3rd Int. Conf. Soil Mechanics and Foundation Engineering (London)., 1, 47–61, 1953.

Spoor, G. & Godwin, R.J. 1978. *An experimental investigation into the deep loosening of soil by rigid tines.* Journal of Agricultural Engineering Research, 23, 243–258.

Spoor, G. "Alleviation of Soil Compaction: Requirements, Equipment and Techniques." Soil Use and Management 22 (2006):113-122.

Spoor, G., Tijink, F.G.J. & Weisskopf, P. 2003. *Subsoil compaction: risk, avoidance, identification and alleviation.* Soil and Tillage Research, 73, 175–182.

Stevenson, F.J. Origin and Distribution of N in Soil. In F.J. Stevenson (ed.) Nitrogen in Agricultural Soils. 1982. American Society of Agronomy, Madison, WI.

Terzaghi, K., Peck, R.B., and Mesri, G (1996). Soil Mechanics in Engineering Practice, 3rd Edition, John Wiley & Sons, New York.

Tisdale, S. L. and W. L. Nelson, Soil Fertility and Fertilizers, The MacMillan Co., 1966.

Tuncer, E.R. and Lohnes, R.A., 1977, An engineering classification for basalt-derived lateritic soils., Eng. Geol., 4, 319– 339.

Tyner, J.S., Wright, W.C., and Dobbs, P.A. (2009). *Increasing exfiltration from pervious concrete and temperature monitoring.* Journal of Environmental Management. 90, 2636-2641.

Underground Soil and Thermal Conductivity Materials Based Heat Reduction for Energy Efficient Building in Tropical Environment. Indoor and Built Environment 24(2):185-200. DOI: 10.1177/1420326X13507591. lam, M.R., M.F. Zain, A.B. Kaish, A. Jamil. 2015.

United States Geological Survey (2004). Landslide Types and Processes. Fact Sheet 2004-3072.

Urban, James. 2008. *Up by Roots*. International Society of Arboriculture: Champaign, IL.

Varnes, D.J. (1978). Slope movement types and processes. In: Special Report 176: Landslides: Analysis and Control (Eds: Schuster, R. L. & Krizek, R. J.). Transportation and Road Research Board, National Academy of Science, Washington D. C., 11-33.

Vieux, Baxter. *Distributed Hydologic Modeling Using GIS*. Dordrecht, Netherlands: Kluwer Academic Publishers, 2001.

Virginia Tech. 2012. Soil Profile Rebuilding Specification. Departments of Forest Resources and Environmental Conservation, Horticulture, and Crop and Soil Environmental Sciences.

WG/WLI (1994). A suggested method for reporting landslide causes. Bull. Int. Assoc. Eng. Geol. 50 (1), 71e74.

Wraith, J.M., and R.J. Hanks. 1992. Soil thermal regime influence on water use and yield under variable irrigation. Agron. J. 84:529-536.

Yagiz, S., 2001, Brief note on the influence of shape and percentage of gravel on the shear strength of sand and gravel mixture., Bull. Eng. Geol. Environ., 60(4), 321-323.

Yardým, Y. and Mustafaraj, E., 2015, Effects of soil settlement and deformed geometry on a historical structure, Nat. Hazards Earth Syst. Sci., 15, 1051–1059.

Youdeowei, P.O. and Nwankwoala, H.O., 2013, Suitability of soils as bearing media at a freshwater swamp terrain in the Niger Delta., J. Geol. Min. Res., 5(3), 58 – 64.

INDEX

www.ingramcontent.com/pod-product-compliance
Lightning Source LLC
Chambersburg PA
CBHW082003190326
41458CB00010B/3058

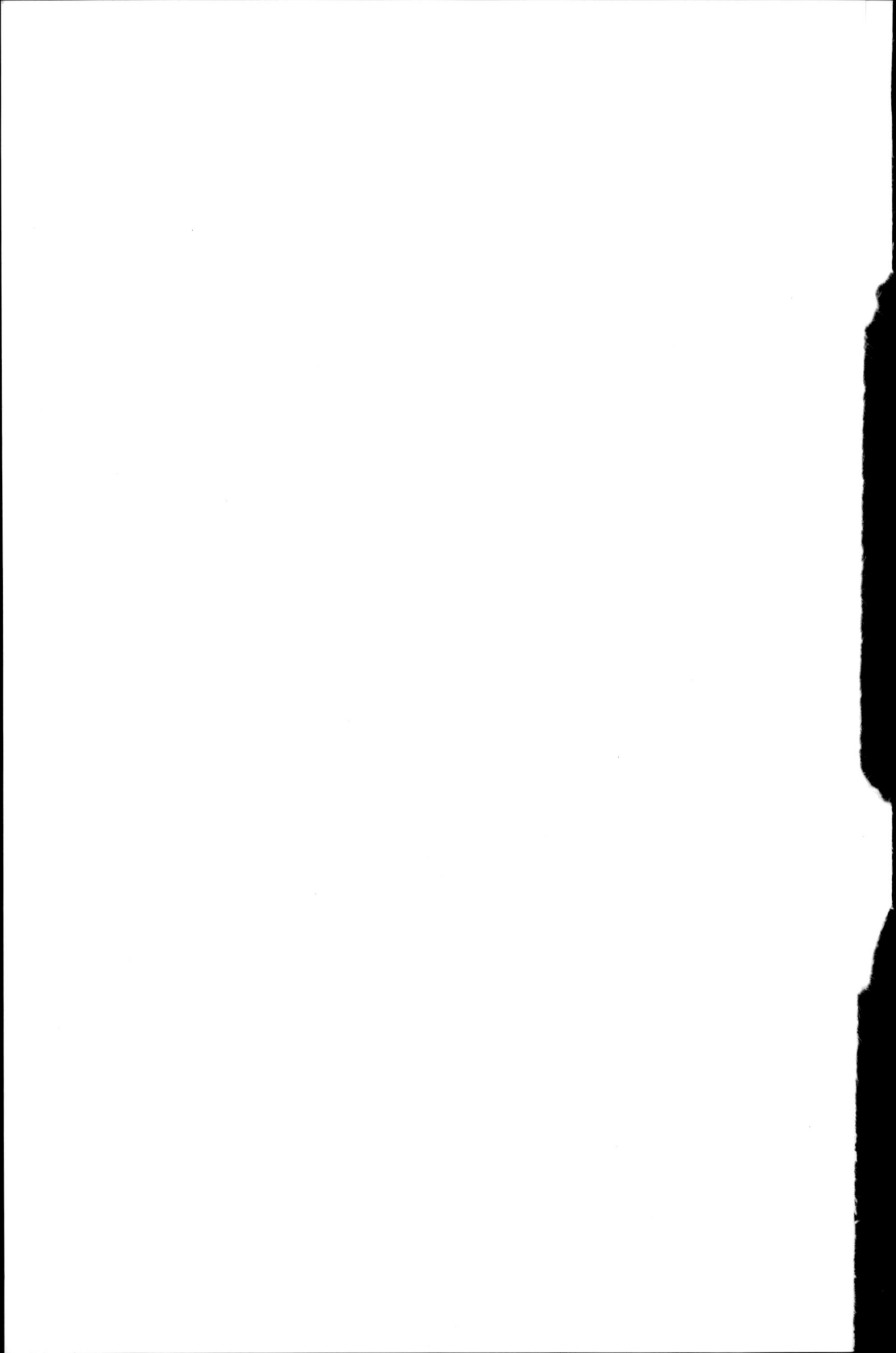